Animals of the African Year
the ecology of East Africa

Animals of the African Year
the ecology of East Africa
by Jane Burton

PHOTOGRAPHS BY JANE BURTON

BOOK CLUB ASSOCIATES
LONDON

Previous page:
Grant's gazelles graze lush pasture during the wet season,
while rain almost blots out the distant hills.

A lion rests in the shade of a bush, guarding his nearby kill.
Uncomfortable in the midday heat, he is plagued by flies and
is panting heavily to keep cool.

The author gratefully acknowledges references
to the many authors whose papers on various aspects of
East African ecology have been published in the
East African Wildlife Journal, volumes I–IX.
Thanks are also due to the National Museum of Kenya, Nairobi
and to the British Museum of Natural History.

Index compiled by Jacqueline Pinhey
Photographer's agent: Bruce Coleman Ltd.

This edition published 1979 by
BOOK CLUB ASSOCIATES
by arrangement with Eurobook Limited

Printed in Italy by New Interlitho Spa

Introduction

East Africa is one of the most exciting places on earth. It contains the highest mountain in the continent of Africa, and the largest and deepest lakes. It is slashed across by the two long gashes of the spectacular Great Rift, pocked by hundreds of volcanoes, many small and long since dead, a few of immense size and beauty, and some still active today. Its old lava flows, its soda lakes, its sheer precipices and granite domes tell of the turmoil which stretched and tore and remodelled the earth's crust here, and produced some of the most spectacular and beautiful scenery in Africa.

The greater part of the area lies within five degrees north and south of the equator. There are no winters or summers, but two annual rainy seasons and two dry ones; and the day length scarcely changes throughout the year. On the equator the rains may be evenly spaced, separated only by mild droughts or by no drought at all. The countryside remains green, plants may flower or seed at any time, animals breed the year round. At the other extreme the drought may be severe and may last for the greater part of the year. Here the few storms of the annual rainy season produce fresh, spring-like greenery full of nesting birds, baby mammals, insects and flowers; but autumn comes almost before there has been any summer, and the wintry, apparent-deadness of leafless twigs and dry dust returns to the land.

The great variety of terrain of East Africa, combined with the varied climate, results in a large number of distinct habitats in which animals can live. These range from the parched, inhospitable deserts of northern Kenya through thorn scrub, rolling grassland, and many forms of tree savannah to the dripping forests and rain-soaked moorlands of the mountains. There are even patches of lowland rain forest in the extreme west of the area. In some of these habitats the profusion of animal life is more rich and varied than almost anywhere else on earth. Not only are there many species and vast numbers of large mammals, but there is a multiplicity of smaller life forms, each adapted to fill a place within its particular natural community.

Not long ago, most of East Africa was an untouched wilderness, teeming with this marvellous abundance of wildlife. After the remarkable killing-spree in which man indulged during the last century much of the wilderness and its wildlife no longer exists.

The dominating factor in the ecology of the area has been the activities of man, but it is not the purpose of this book to dwell on the changes he has brought about, though mention of human influence is unavoidable. Fortunately, enough of the wilderness survives to bear witness to the interlocking patterns of animal and plant relationships that have developed over many thousands of years. The complexity of these relationships is such that detailed coverage of the ecology of even one habitat might take a whole book, while adequate coverage of the whole of East Africa is a near impossibility. Instead of attempting to unravel the whole intricate web of life throughout the area, this book shows typical animals and plants interacting with one another, yet fitting perfectly into their chosen habitats, surviving the rigours of the yearly cycle of events in the climates in which they live. I hope thus to convey something of the over-riding unity and sense of purpose that is so evident in natural communities, unspoiled by civilized man.

Jane Burton
Albury
Surrey

The East African Region

flamingos

Australopithecus

Deinotherium

Chalicotherium

giant baboon

Sivatherium

marabou stork

Africa, with its great range of climate and vegetation—from the Mediterranean fauna and flora in the north, through the perpetual green of the great forests, to the extreme conditions of life on the high mountains or in the deserts—has so many habitats and a fauna of such great interest and variety that it defies more than a superficial glance in a small space. It is convenient, therefore, to take a corner of the continent which is fairly representative of much of the rest of Africa, in the hope that a longer look at its ecology will be rewarded by a greater understanding of the continent as a whole. For East Africa has forests as well as grassland, and deserts as well as mountains, and the ranges of many of its most spectacular animals extend well beyond the political boundaries of the area.

East Africa is only a small part of the zoogeographical region known as the Ethiopian, which comprises the whole of Africa south of the Sahara. It includes the three large territories of Kenya, Uganda and Tanzania together with Rwanda and Burundi, and is bounded for the most part by natural barriers. To the east lies the Indian Ocean, to the west the lakes and mountains of the Western Rift. To the north is Ethiopia, a highland area with a somewhat different fauna, separated from Kenya by stony deserts. The Sudan and the Somali Republic, though separate, merge ecologically and geographically into northern Uganda and northern and eastern Kenya. To the south, Tanzania is clearly cut off from southern Zaire and from Malawi by Lakes Tanganyika and Malawi, but its boundary with Zambia is clearer on a map than on the ground. Between Tanzania and Mozambique the demarcation line is the Ruvuma River; but vast areas are geographically similar on either side of the river.

It is convenient to divide the continent of Africa roughly into quarters: North Africa, which is characterized by deserts; West and Central Africa, with their equatorial forests; South Africa which consists largely of high plateaux covered with dry grassland but also contains an area of desert; and East Africa, which, except for a narrow coastal plain, is mainly highland and is characterized by sub-desert steppes.

The distribution of the main vegetational zones neatly coincides with the distribution of rainfall throughout the continent. The area of highest rainfall is in West and Central Africa, and produces the great tropical forests of the Congo and Niger Basins. Around the forest edges the vegetation grades into woodland, into wooded savannahs and open grasslands in areas of seasonal rain, and finally into desert in the areas where rainfall is most sparse. The deserts also coincide with more temperate conditions in the northern and southern quarters of the continent, where hot summers alternate with warm winters. In the middle part of Africa it is hot all the year round, although in the highlands nights may be cool or even cold.

The history of exploration

A hundred years ago the continent of Africa held many secrets and challenged the most intrepid explorers. The Mountains of the Moon and the source of the Nile remained unknown to Europeans. Yet the discovery of Africa had begun about fifty centuries earlier when the Egyptians sent caravans southwards to bring back ebony, ivory and slaves to their ancient capital, Memphis. During the time of the Pharoahs the cultural influence of Egypt on the interior of Africa was very great and it can still be discerned today. Later the Phoenicians, travelling far beyond the Mediterranean in search of trade, sailed south along the Atlantic coast of Africa. About 600 BC a fleet took three years to sail around the continent, and in 450 BC Hanno took a vast fleet of sixty Phoenician galleys and 30,000 prospective settlers down the west coast almost to the equator. After the Phoenicians came the Greeks, followed by the Romans, who established their empire in North Africa and pushed southwards into Ethiopia and beyond. There was also some contact between China and Africa a thousand years ago. By that time Arab traders were well established on the east coast. They travelled to and fro on the monsoon winds, carrying in their dhows cloth, beads and ironware which they exchanged for ivory, slaves and raw metals. The Arabs rarely penetrated inland, relying on African traders to bring the goods and slaves to them, but large settlements grew up along the coast, and the Arabs

Previous page: Two million years ago East Africa was inhabited by a profusion of animals, many of giant size. Some – flamingos and marabous, zebras, antelopes and hyenas – had already evolved into much their present form. Some, such as rhinoceroses and baboons, were giants; their descendants, though much like them in form, are dwarfs in comparison. Others, then at the dawn of their history, were small in comparison with later forms: *Australopithecus*, with his ape-like jaws and small brain, was much shorter than modern Man. Other animals such as the chalicotheres, four-horned *Sivatherium*, the sabre-toothed tiger, and elephant-like *Deinotherium* were coming to the end of their span and have vanished entirely from the scene during the intervening millenia.

The clawed frog, a faunal link with South America and with the ancient supercontinent of Pangaea of which Africa once formed a major part.

were undisputed masters of East Africa until the arrival of the Portuguese in the fifteenth century.

It was mainly the lure of West African gold that first drew the Portuguese to explore Africa, though they also hoped to wrest the trade monopoly from the Arabs. During the first half of the fifteenth century they inched warily down the West African coast under the direction of Prince Henry the Navigator. Fifty years later they had established a safe harbour at Lagos, and with it the first of a chain of coastal forts intended to deter other European powers. Vasco da Gama's discovery of the Cape sea route to the Far East gave Portugal a monopoly in trade in West Africa, though the Arabs remained strong in the East. Portuguese strength was challenged in the seventeenth century, however, when other European nations, England, Holland, France, Sweden and Prussia, became interested in the profitable slave trade. By the end of the century the Portuguese had lost their brief supremacy and England had established hers.

Some twenty million Africans were deported as slaves and a further ten million died before em-

barking or on passage to the Americas, before opposition to slavery, started by the Quakers, gained momentum. As the era of slavery neared its end, people began to take a wider interest in Africa. The industrial revolution created a demand both for new sources of raw materials—such as rubber—and for new markets for its products. Scientific curiosity was aroused. James Bruce's explorations in Ethiopia encouraged the founding in 1788 in London of the "Association for promoting the Discovery of the Interior Parts of Africa". For although the Romans had colonized North Africa and the coasts had long been settled, the vast centre of Africa was unexplored. The obstacles were formidable: there were the natural barriers of deserts, swamps and mountains, as well as the hazards of disease and wild animals. The inhabitants were justifiably hostile, fearing strangers as harbingers of slavery.

In 1795 the first of the great explorers of Africa, Mungo Park, survived an exceedingly hazardous journey to find the River Niger. In 1822 Hugh Clapperton became the first white man to enter the

15

great northern Nigerian market city of Kano. The Frenchman, René Caillie, was the first European to visit the legendary and ancient Timbuktu and return to tell the tale, and Heinrich Barth of Hamburg alone survived a three-man expedition to the region between Timbuktu and Cameroon. Thus France and Germany as well as Britain were drawn into the scientific exploration of Africa. At the same time the great wave of missionary exploration got under way, reaching its peak in the work of David Livingstone.

During the second half of the nineteenth century East and Central Africa were gradually explored by Europeans. Richard Burton and John Speke reached Lake Tanganyika in 1857; a few years later Speke, with James Grant, discovered the source of the Nile—the Ripon Falls, where the Nile leaves Lake Victoria. In the early 1880s Joseph Thomson who, like Grant, has a gazelle named after him, led a Royal Geographical Society expedition to East Africa. He climbed Mount Kenya—whose existence was still questioned—and travelled down the Rift Valley to Uganda.

By 1900 the heart of Africa was no longer a complete mystery. It had been explored and mapped, but still remained challenging to explorers, adventurers and missionaries. All over Africa the great European powers were scrambling to stake their claims to the continent in the final decades of the last century. The apportioning was a direct result of the discoveries of the previous half century, explorers creating spheres of influence for their home countries. Often the divisions established had little connection with the boundaries of indigenous tribes. Borders were marked on maps along latitudes and longitudes, or following mountain ranges, lakes or rivers—hence the neatness of East Africa on the map.

When the countries of Europe began to rule Africa, European–African relationships were transformed. Anthropologists intensified studies of African societies and customs, and the Africans, brought daily into contact with white settlers and administrators, began to adopt European customs to varying extents. Some tribes still resist the alien clothes, foods and customs; others have adopted European ways. High-speed jet airliners now deposit European tourists in parts of Africa which a hundred years ago could only have been reached after months of death-defying effort. Yet in another sense Africa is still the Dark Continent, and the end of African discovery is nowhere in sight.

In the twentieth century we know as much of the ancient cultures of Africa as the nineteenth century knew of its geography. No form of writing was invented in Africa, south of the Sahara, until very recently, so that all traditions are oral. In South Africa, spoken histories have been traced back over a thousand years, but archaeology offers the best opportunity to uncover the past, and archaeology in most of Africa is still in its infancy. Egypt has attracted the world's greatest archaeologists but whole areas in other parts of Africa have scarcely been touched.

From the writings of Arab explorers we know something of the grandeur and power of the ancient kingdom of Ghana and of the fabulous wealth of the kingdom of Mali which succeeded it ten centuries ago; but of the great civilization of Kush or Nubia we know almost nothing. The written Kushite language remains undeciphered, their temples, palaces and tombs undug. Descendants of the Kushites may have built the stone towns and terraces whose ruins can still be seen in East Africa, near Lake Natron and elsewhere, but the archaeological evidence here, as for most of the continent, is fragmentary and inconclusive.

The dawn of man in Africa

The remote ancestors of man in Africa have received more detailed study than his immediate ancestors. Africa may well have been the cradle of mankind, for pre-humans, the australopithecines, lived in southern Africa about two million years ago, at the dawn of the Pleistocene Period. Fossil remains of these ape-men have been found in South Africa, at Olduvai Gorge in the Serengeti Plains of north Tanzania, and at Baringo and Rudolf in Kenya. The large australopithecine *Zinjanthropus*, nicknamed Nut-cracker Man because of its small brain and very large and heavy jaws, appears to have lived in East Africa contemporaneously with another hominid, named by the finder, Dr L. S. B. Leakey, *Homo habilis*. It seems to have been a small, light plains-dweller, while Nut-cracker Man, larger and more powerful, was possibly a forest-dweller. Controversy exists over the validity of separating these ape-men and true men into the genera *Australopithecus* and *Homo*. It is apparent that both walked with an erect posture and that *Homo habilis* made and used crude tools. East Africa is at the centre of the area in which very early pebble tools have been found, suggesting that the earliest human culture may have arisen here.

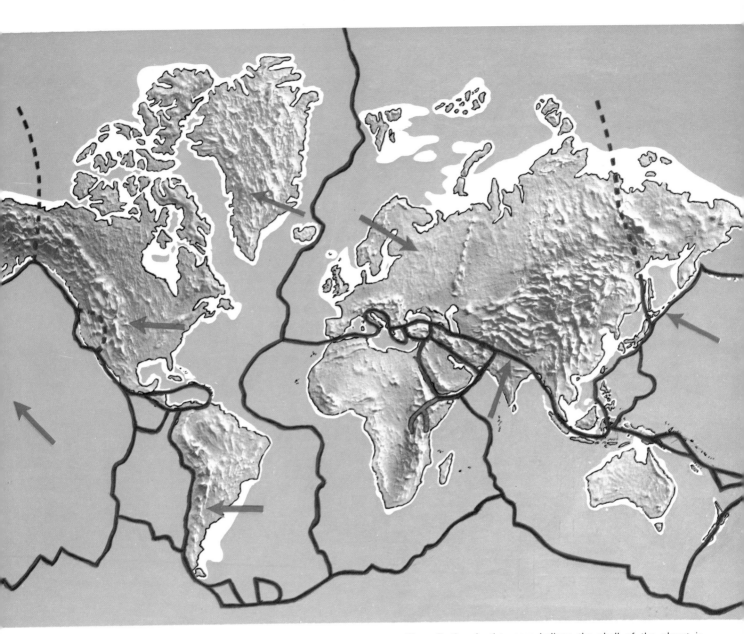

Many Earth scientists now believe the shell of the planet is broken into "plates", which are moving gradually in different directions as indicated by the red arrows. At the boundaries of plates, pressure has caused great mountain ranges to be thrown up – for example the Himalayas, the Andes and the Rocky Mountains. The Great Rift, shown here in red, is part of a series of submarine rifts along the shore of the Indian Ocean. Over five thousand kilometres long, it runs from south of the Zambezi Delta through East Africa and Ethiopia to the Red Sea and beyond. Two branches of the system run through East Africa, the Great Rift itself through Kenya and Tanzania, and the Western Rift to the west of Lake Victoria, through Rwanda and Burundi.

East African australopithecine and habiline man lived around the edges of lakes in temporary encampments. Their living floors are littered with bone fragments of fish, reptiles, birds and mammals as well as with naturally and artificially fractured quartz and obsidian stones. Outside the living area there are quantities of split and hammered bones of big mammals. The early man-like habilines probably lived partly by hunting small or slow creatures such as rodents, fish, snakes, frogs and tortoises, and partly by scavenging from predator kills. The australopithecines probably had a more vegetarian diet, like the modern gorilla, with whom they share certain skull characteristics. They probably ate bulbs, fruits, stems and nuts, as well as a small amount of meat, which they may have killed for themselves using bone or horn weapons.

Early man's animal neighbours

In the Pleistocene lake beds at Olduvai and at Olorgesailie near Lake Magadi, Kenya, an amazing variety of fossil animals contemporary with the hominids have been found. Many of the herbivores were enormous. For example, there was a giant sheep with a horn span of over a metre, and a pig as big as the modern rhinoceros. There was the elephant-like *Deinotherium*, with downward-curving tusks in the lower jaw and the tall straight-tusked elephant; there were the hefty, digging and browsing chalicotheres; and the elk-like *Sivatherium*, a relative of the giraffe, with a massive body over two metres long and two pairs of horns. Ostriches attained an even greater size than they do today, and there was a giant species of baboon and giant tortoises. There were various ancestral antelopes and gazelles, a big zebra and the little three-toed ancestral horse *Hipparion*, as well as vultures, flamingos, rhinoceroses and many other animals similar to those found in East Africa today. Preying on the herbivores were hyenas, the sabre-toothed tiger with its 18 to 20 centimetre fangs, and other carnivores of various sizes down to bat-eared foxes and genets.

The immense variety of African mammals then and now makes it surprising how little is known of their early history. Sedimentary deposits of the right age, in which remains might be found, are scarce in Africa. The earliest mammalian ancestors of today's fauna appear in lake beds in Egypt, dating from Eocene and Oligocene times, 50 to 30 million years ago, where the first diminutive elephants and the first monkeys and apes have been found. But south of the Sahara there are few fossil-bearing sediments earlier than the Miocene, and even these are rare except in East Africa. Here, however, on sites on the islands of Lake Victoria and near its shores in western Kenya and eastern Uganda, hundreds of fossil mammals that lived from 15 to 20 million years ago have been found, enabling palaeontologists to build up a picture of the fauna of the time.

Primates, normally rare among fossils, are exceptionally abundant here. There were three species of *Proconsul* apes, the largest as large as a gorilla and almost certainly related to it; an ancestral gibbon, and at least two galagos. Monkeys were rare at this time. Several kinds of insectivores have been found; these, too, are very rare from the Miocene elsewhere. Rodents were abundant, especially *Diamantomys*, an animal like a cane-rat, previously known only from the diamond deposits of South-West Africa. *Anthracotheres*, or ancestral hippopotamuses, and ancestral giraffes suggest that both these animals originated in Africa during the late Miocene. At that time there were no true antelopes—only chevrotain-like tragulids—but *Deinotherium* was already present, as were mastodonts. Carnivores were rare, but there were a great many reptiles—snakes, turtles, lizards and crocodiles.

Before the mammals

If mammalian ancestors are scarce, the reptile-like ancestors of the mammals are abundant in East African fossil-bearing rocks. The fauna of the East African Upper Permian is similar to the fauna of other parts of Africa at that time. Fossils of this age have been found in the Ruhuhu Valley near Lake Malawi in south-western Tanzania. In the same area fossils have been found from the Middle Triassic, about 210 million years ago—fossils which are of very great importance in furthering our knowledge of the prehistoric reptile fauna of the whole of Africa, because although abundant fossils have been found for earlier and later periods in South Africa, there is a gap in the sequence during the Middle Triassic, when conditions there were unfavourable for the laying down of fossiliferous strata. The Manda fauna of the Ruhuhu Valley neatly fills this gap. Here were reptiles of four main groups, including the crocodile-like ancestors of dinosaurs and of flying reptiles, crocodiles and birds, and the mammal-like reptiles from which true mammals evolved.

By the end of the Triassic the first dinosaurs were beginning to appear, and with them the first true mammals—small, shrew-like creatures that probably rummaged about among the leaf-litter for insects much as modern shrews do today. Gradually the dinosaurs increased in size until during the Upper Jurassic, 150 million years ago, the Earth was dominated by these gigantic animals. The biggest East African dinosaur was *Brachiosaurus*—indeed it was the largest land animal the world has ever known, weighing twenty times as much as a modern bull elephant, and able to browse up to 14 metres above the ground. Together with other giant herbivorous saurians, such as the plated stegosaurs, it must have demolished plants by the hundredweight, and was itself preyed upon by giant carnosaurs.

The fossil reptiles of East Africa have been satisfactorily documented, but so far there has been little to show of the fossil amphibians which gave rise to them. However, the sequence between the amphibians and the fishes is more completely documented in the fossil record than is the sequence between any other large groups. The ancestors of the four-legged land vertebrates were fleshy-finned fishes such as *Eusthenopteron*, of North America and Europe, which flourished in Upper Devonian times 350 million years ago. These fishes had primitive lungs and could crawl over land; they probably gave rise to the animals with well-developed limbs but fish-like tails such as *Ichthyostega*, the earliest known amphibian. The only surviving fleshy-fins today are the coelocanth of the Indian Ocean and the three genera of freshwater lungfishes. Lungfishes (page 80) are not in the direct line of descent between the rest of the fishes and the amphibians, but represent the kind of fish—capable of moving onto land and breathing by lungs—that led to the evolution of the amphibians.

The continent

Geophysicists have known for a long time that the Earth's continents consist of materials lighter than those composing the deeper layers of the Earth, and are floating upon them. The most favoured and acceptable theory about how the continents came to occupy their present positions is Wegener's theory of Continental Drift. Wegener pointed out that the continental shelves of the opposing continents could be fitted together very accurately. For example, the eastern bulge of South America could be made to fit neatly into the large bight of western Africa, as if the two had once been joined and had drifted apart. There is evidence, too, that Australia was earlier joined to south-east Africa. This theory of Continental Drift has now been confirmed along several lines of research.

As far as the animals are concerned, indications of continents drifting apart can be found in the distribution of several groups. One of the most spectacular is that of the lungfishes. The three lungfish genera are found today only in Australia, tropical Africa and tropical South America: if the theory is correct, in the heyday of the lungfish, these areas were more or less contiguous.

Other animals sharing this distribution are the large running birds, or Ratites: the emu and cassowaries of Australasia, the ostrich of Africa, and the rhea of South America. The African clawed frog has its South American counterpart in the pancake-shaped Surinam toad, while the side-necked terrapins, whose necks bend laterally in the horizontal plane, are also restricted to the southern continents: the snake-necked turtle in Australia, the helmeted water tortoise in Africa and the matamata in South America. Among the lower animals the odd-looking *Peripatus*, half worm, half insect, has a similar distribution today, and although it is no longer regarded as a direct link between worms and insects, as was once supposed, it nevertheless represents an archaic type in the lineage of insects.

The theory of Continental Drift is now generally accepted. Earth scientists are also confident that the continents are drifting because the surface of the Earth is broken up into a number of rigid plates of different sizes, all moving in different directions carrying the continents on their backs. What propels the plates themselves is still a matter for debate: convection currents welling up from the hotter underlying mantle of the earth may be moving them, or one plate becoming cold and slipping beneath another may cause currents that drag other plates to take its place. When plates collide, the great mountains of the world are thrown up. India colliding with Asia threw up the Himalayas, while the ocean that lay between them vanished. Similarly, the Urals were formed when Europe and Asia were in collision. But the mountains of East Africa were not created by the convergence of plates, for they do not occur at plate boundaries. They are paradoxical mountains that rose on huge domes, rather like gigantic bubbles welling up when heated from below, to form the great peaks and ranges that can be seen throughout the region today.

The glistening white cone of Kibo, highest of the twin summits of Mount Kilimanjaro, Africa's highest mountain. As the snowfields and glaciers catch the sunset, a Masai giraffe, already in darkness, is silhouetted against the lower slopes.

The Habitats

The Mountains

East Africa is one of the great volcanic areas of the world. Most of its higher mountain peaks are the cones or craters of extinct volcanoes. Some, such as Mount Elgon, and Mount Longonot in Kenya are still steaming. Oldonyo Lengai in southern Tanzania is gently active, occasionally spewing out soda lava so that its cone looks snow-capped. Astride the international boundaries of Rwanda, Uganda and Zaire the Virunga volcanoes are still violently active.

Almost wherever we look we are reminded of the violent vulcanism of the past: by the volcanoes themselves, by basalt cliffs and still-naked lava flows, by boulders of pumice and obsidian, by soda lakes and the contorted strata of isolated hills; while steam jets and boiling springs show that volcanic activity seethes not far below the surface even today.

There are two main theories to explain why this extensive volcanic activity occurred in East Africa. Earth scientists are now confident that the surface of the Earth is broken up into a number of rigid plates, all moving in different directions, carrying the continents. At the boundaries between the plates major geological events take place and many of the world's volcanic areas are there. Other volcanic areas, of which East Africa is one, occur inland from plate boundaries. They may be either along old fault lines or weaknesses in the crust; or they may occur over *hot spots*, strong sources of volcanic energy stationary beneath the plates, in about twenty different parts of the globe. According to the second theory the African plate may have passed over one of these hot spots, which heaved up the north-eastern quarter of the continent into huge domes. Directly over the hot spot, volcanoes punched their way through to the surface, pouring out molten rocks from the Earth's mantle; and extensive rifts split the surface of the land. As the plate continued imperceptibly on its way, propelled by the same energy source, the first volcanoes were carried away from the hot spot and they cooled, while others sprang up violently to take their place.

The only violent volcanoes in East Africa today are two of the Virunga peaks, Nyamlagira and Nyiragongo. The most violent is Nyamlagira. The scene inside its mile-wide crater is one of utter desolation: black solidified lava encrusted with sulphur, sulphurous vapours steaming from clefts

On cool, dewy mornings small steam jets can be seen gently puffing from vents in the Rift Valley wall. Later, as the air heats up, the steam vaporizes and the jets are invisible. In places, whole hillsides appear to be smoking in the early morning sunlight.

A cold mountain stream cascades down the face of a basalt precipice in the Aberdare Mountains. Basalt is lava which cooled rapidly and, while solidifying, formed into characteristic hexagonal columns. Redwinged starlings, birds of the mountains, nest in coolness and safety on a ledge behind the fall. Giant groundsels sprout from the cliff.

and smoke-holes, and two great cones of volcanic ash belching out smoke. Every few years the mountain roars and spews out a scarlet river of lava which rushes down the slopes at 12 kilometres an hour, sweeping aside all plant life until it plunges hissing into Lake Kivu, 15 kilometres away. During the eruption elephants leave the district in terror, but small animals are attracted to the fires at night, and predators in turn are attracted to the swarms of insects and rodents, dodging after them through the hail of rocks and ashes.

The neighbouring and somewhat higher peak, Nyiragongo, also becomes active at the same time. This is one of the few volcanoes in the world that has a liquid lava lake inside its crater, and the lake boils and seethes when Nyamlagira erupts. The

Mount Kilimanjaro, too, is still gently puffing steam. At almost 6,000 metres it is the highest mountain in Africa today. Elgon has no permanent snow, but Kibo, the highest peak of Kilimanjaro, is covered with snowfields and glaciers. Mount Kenya also bears glaciers and snowfields. This mountain is unique, for its single peak is a solid lava plug of an extinct volcano. The ancient crater rim has been almost eroded away, except for some fantastically-shaped remnants on the outlying shoulders. It has been described as the most magnificent example of mountain architecture, a single peak standing 600 metres above its gently sloping pedestal of forest and moorland.

In the line of the Western Rift between Lakes Edward and Albert is another great mountain range, the Ruwenzori, the fabled Mountains of the Moon. Unlike most other East African mountains it is not volcanic, but is a fault block of very ancient greenstone heaved up from the basement rock which underlies the rest of the continent. The 3,300 metre Cherangani, an escarpment in the Great Rift, is another non-volcanic range. The Ruwenzori range is a high, domed plateau from which the peaks jut up in groups, the six central ones sparkling with perpetual snows and glaciers, the tallest, Mount Stanley, rising to 5,000 metres. Below the snow-line are numerous lesser peaks; the whole range is immensely rugged and nearly always shrouded in cloud. Ruwenzori has the highest rainfall of any East African mountain; great rivers gouge deep gorges as they hurtle down towards Lake Albert and the Nile. The exceptional humidity and the absence of a dry season gives year-round plant growth so that the upper slopes are clothed in most luxurious cloud-forest.

Kenya, Kilimanjaro and Ruwenzori are the only African mountains to bear permanent ice, for the Atlas in Morocco, although snow-covered, has no glaciers. Kilimanjaro still has the most extensive glaciers of the three, but they are steadily shrinking, as are glaciers elsewhere in the world, because rainfall no longer exceeds the rate of evaporation. The Ruwenzori glaciers are ice-caps, rather than ice-rivers; great rounded cornices covering the high cols and peaks, with huge icicles underpinning the exposed ridges. Today they do not extend below 4,200 metres, but at one time huge ice-caps covered all the East African mountains down to 1,500 metres, and great rivers of ice flowed down into the valleys. The retreat of the East African glaciers may be due to the world-wide tendency towards a warmer, drier climate.

other peaks are dormant now; Mikeno and the pointed Karisimbi, both over 4,000 metres high; flat-topped Visoke; Sabinio, Muhavura and Mgahinga; all were once explosive but are quiet now.

The three great isolated peaks of East Africa, Kenya, Kilimanjaro and Elgon, are huge extinct volcanoes. Mount Elgon may once have been the tallest of the three—it has the greatest circumference at the base—but today it is gently sloping and only 4,200 metres high. Its crater is enormous, the second largest in the world; only the spectacular Ngorongoro caldera in Tanzania is bigger. The floor of the Ngorongoro is grassland teeming with big game, but cold and altitude prevent more than a few specialized animals from browsing the alpine flora of Elgon's crater floor.

Small montane forest trees thin out at about 3,000 metres. A mountain river, dashing downwards over old lava flows, passes a patch of falling bamboos that have flowered and died.

The flame lily climbs among creepers and bushes in clearings and at the forest edge, its stems anchored by tendrilled leaf tips.

Right: East Africa includes three large countries, Kenya, Tanzania and Uganda, and two smaller ones, Rwanda and Burundi. The map shows these, with the mountains and lakes of the region and the areas set aside for wildlife as National Parks and Game Reserves.

The vegetational zones on East African mountains, from forest round the foot to snow on the highest summits.

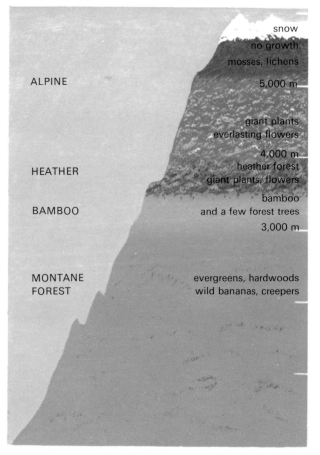

ALPINE

snow
no growth
mosses, lichens

5,000 m

giant plants
everlasting flowers

4,000 m

HEATHER

heather forest
giant plants, flowers

BAMBOO

bamboo
and a few forest trees

3,000 m

MONTANE
FOREST

evergreens, hardwoods
wild bananas, creepers

The forests There are differences in rainfall on East African mountains but because of similarities of altitude, low temperatures and high radiation there are very similar belts of vegetation. The lower slopes, up to a height of 2,400 or 3,000 metres, are encircled by rain forests called *montane* forests. Montane forest is very similar to true tropical forest but has smaller trees, fewer lianas and several gymnosperms—non-flowering evergreen trees. Stands of mature trees, of camphor, *Podocarpus*, cedar and red stinkwood, tower 30 to 40 metres above a wide variety of understorey trees. Beneath these great trees the ground may be clear but for a thick layer of leaf-mould and some small herbs. In other places, extensive thickets of wild bananas or bracken grow. The trees are festooned with parasitic creepers, beard lichens and tree orchids. Dwarf Kikuyu grass and pleasant flowering plants grow in the glades, together with groves of unpleasant giant stinging nettles.

At one time most of East Africa was covered by such forests, but today only about two per cent of the forest remains, at least in Kenya and Tanzania. Fragments of forest are scattered over most of Africa, and many of these remnant patches contain a remarkably similar fauna. High altitude butter-

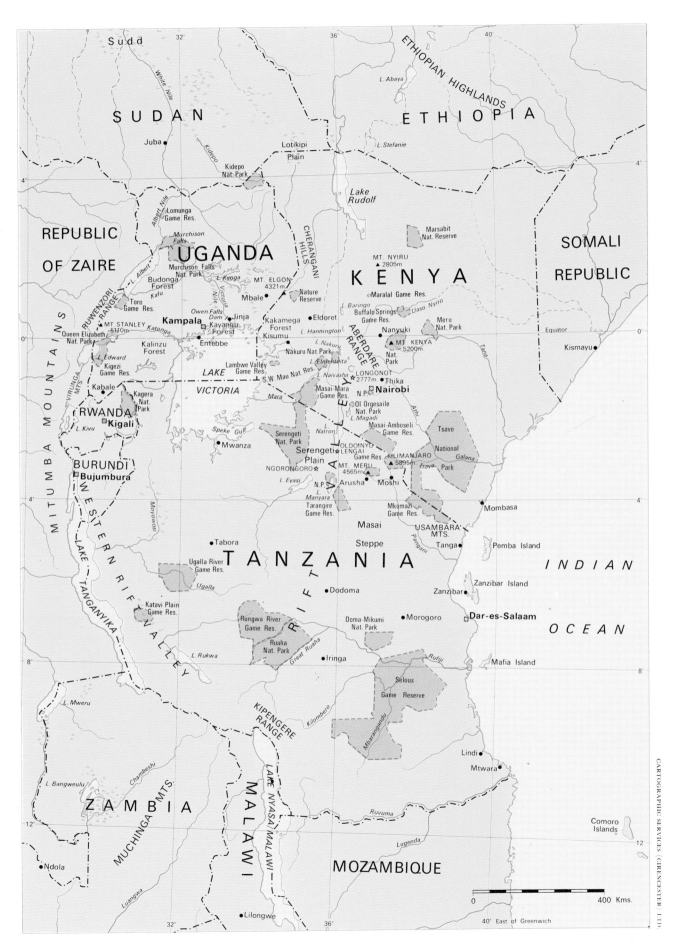

Sudd

32°

SUDAN

White Nile

L. Abaya

ETHIOPIAN HIGHLANDS

36°

40°

ETHIOPIA

4°

Juba •

Kidepo

Lotikipi
Plain

L. Stefanie

Kidepo
Nat. Park

*Lake
Rudolf*

SOMALI

4°

REPUBLIC
OF ZAIRE

Lomunga
Game Res.

*Murchison
Falls*

UGANDA

Albert Nile

L. Albert

Murchison Falls
Nat. Park

Budonga
Forest

L. Kyoga

Kafu

CHERANGANI
HILLS

MT. ELGON
4321m

Mbale •

Nature
Reserve

Marsabit
Nat. Reserve

MT. NYIRU
▲ 2805m

KENYA

REPUBLIC

Toro
Game Res.

RUWENZORI
RANGE

▲ MT. STANLEY
5110m

Kampala ⊡

*Victoria
Nile*

Kakamega
Forest

Kisumu •

Kayango
Forest

Eldoret •

Buffalo Springs
Game Res.

Maralal Game Res. •

L. Baringo

Uaso Nyiro

Meru
Nat. Park

Nanyuki •

0°

Queen Elizabeth
Nat. Park

Katonga

Jinja •

Owen Falls
Dam

Entebbe •

L. Hannington

ABERDARE
RANGE

▲ MT. KENYA
5200m
Nat.
Park

Equator

Kismayu •

0°

MITUMBA MOUNTAINS

L. Edward

Kigezi
Game Res.

VIRUNGA
MTS.

Kabale •

Kalinzu
Forest

Kagera
Nat.
Park

L. Kivu

RWANDA

Kigali ⊡

LAKE

VICTORIA

Lambwe Valley
Game Res.

S.W. Mau Nat. Park

L. Nakuru

Nakuru Nat. Park

L. Elmenteita

L. Naivasha

LONGONOT
☆ 2777m

Masai-Mara
Game Res.

Mara

L. Natron

Thika
Nairobi ⊡

N.P.

Ol Orgesaile
Nat. Park

L. Magadi

Masai-Amboseli
Game Res.

Tsavo

National

Athi

Tana

BURUNDI

WESTERN RIFT VALLEY

Bujumbura ⊡

Mwanza •

Speke Gulf

Serengeti
Nat. Park

Serengeti
Plain

NGORONGORO ☆

L. Eyasi

OLDOINYO
LENGAI
Game Res.

▲ KILIMANJARO
5895m

▲ MT. MERU
4565m

Park

Galana

Tsavo

LAKE TANGANYIKA

Moyowosi

N.P.

*L.
Manyara*

Arusha •

Tarangire
Game Res.

Moshi •

Mkomazi
Game Res.

USAMBARA
MTS.

Mombasa •

4°

INDIAN

Tabora •

TANZANIA

Masai

Steppe

Pangani

Tanga •

Pemba Island •

4°

Ugalla River
Game Res.

RIFT

Zanzibar •

Zanzibar Island

OCEAN

Ugalla

Dodoma •

Katavi Plain
Game Res.

Rungwa River
Game Res.

Doma-Mikumi
Nat. Park

Morogoro •

Dar-es-Salaam ⊡

L. Rukwa

Ruaha
Nat. Park

Great Ruaha

Iringa •

Rufiji

Mafia Island •

8°

L. Mweru

KIPENGERE
RANGE

Kilombero

Selous
Game Reserve

8°

MUCHINGA MTS.

L. Bangweulu

Chambeshi

LAKE NYASA/MALAWI

Mbarangandu

Lindi •

Mtwara •

ZAMBIA

MALAWI

Ruvuma

Comoro
Islands

12°

Ndola •

Luangwa

MOZAMBIQUE

Lugenda

Lugenda

12°

Lilongwe •

0

400 Kms.

32°

36°

40° East of Greenwich

flies are closely related to those of lowland forest, while many other kinds of animals found in East African forests are similar to species found in rain forests elsewhere. Some rain forest animals common elsewhere in Africa, such as some forest monkeys and hornbills, are now found only in the very west of Uganda. Here also may occur the tree pangolin, a very accomplished climber with a long prehensile tail; the brush-tailed porcupine, a smaller, spiny, climbing relative of the familiar, quill-covered crested porcupine; and the tiny dwarf galago which hops about in the tangled creepers and low bushes at the base of forest trees. Giant rats are more widespread in forest, but pottos and Fraser's scaly-tail or flying squirrel cannot now be found further east than Kakamega Forest, a tiny pocket of rain forest near Lake Victoria, and the last piece of true lowland forest left in Kenya. Even this remnant is now fast disappearing before the onslaught of the axes and chain saws of commercial forestry; soon the area will become swallowed up by the population pressure of the surrounding agricultural tribes.

Several East African peoples, notably the Kikuyu and Luo, traditionally live by shifting cultivation, clearing areas of the forest for their gardens and then moving on, allowing the luxuriant secondary growth to reinvade their plots and restore fertility. Today there is practically no more forest for such people to shift to when, after several years, the ground becomes unproductive. Many forests have been almost totally destroyed, the great trees turned into charcoal and the ground impoverished by the growing of maize, which gives back nothing to the soil. The people are innocent, but their population continues to increase. Not only the forest animals, but the Wanderobo, a truly forest people who live by hunting small animals and gathering wild honey, have little forest left to live in. Fortunately, many mountain forests are now included in national parks, for they are valuable catchment areas. But even forest reserves are not always sacrosanct from destructive resettlement schemes.

During the day montane forests, undisturbed by human settlement, are cool, dark places, full of the calls of birds and the crashing of monkeys in the upper canopy. An elephant snapping off a branch with a loud report may send a small bright-chestnut duiker scampering across a sunlit glade. There are many species of forest duikers, curious little hunch-backed antelopes with squat features and low head carriage, rather different from the more

familiar bush duiker of open grassland. Forest duikers range from the tiny blue, only 30 centimetres high, to the relatively large and thickset yellow-backed. Even tinier and as secretive as the blue duiker is the suni, a graceful spike-horned forest antelope preyed on by the majestic crowned eagle. Largest of the forest antelopes is the bongo; in spite of its massive double-spiralled horns it moves easily among dense undergrowth. Duiker, suni and bongo are all mainly nocturnal, so they are very rarely seen.

Giant forest hogs live in small numbers on some East African mountains, in family sounders. They feed mainly at night, on grasses and herbs at the forest edge and on leaves and fallen fruit in the depths of the forest. The largest wild pig in Africa, their huge dark hulks may sometimes be mistaken for young buffaloes as they wallow in some forest glade.

In mountain forests there are no true seasons. Small antelopes and other small forest mammals may give birth, and birds nest, at any time of the year. Plants flower and fruit at any time. Insects, such as the huge metallic purple or green chafers that zoom onto forest flowers, are abundant. Rust-coloured shrew-rats find plenty of small invertebrates to eat as they patter, whiffling, over the leaf mould. Bird armies search the trees for

The blue-shouldered robin chat, a small thrush-like bird with a melodious song, lives in the undergrowth of lowland forests. It may join in bird parties which follow safari ants to snap up the insects they flush.

Left: On three of the Virunga volcanoes and in the Kayonza Forest of Uganda gorillas can often be glimpsed, for they are not uncommon where they do occur. They may even in time accept patient humans as their fellows.

A column of safari ants on the march is an awesome sight—and a powerful ecological force. A hundred thousand tiny predatory insects stream in a close mass over the forest floor like some gigantic superpredator, overpowering and bearing along anything that cannot get out of the way.

small insects or follow the columns of safari ants that pass in an apparently unending stream across game paths and glades, under logs and round boulders. As many as 100,000 ants may pass in column, guarded on each flank by soldiers with formidable jaws. Anything that cannot get out of their path is carried along with them, whole or in shreds, from caterpillars to a recently-gorged python. Flycatchers, shrikes, bulbuls, barbets, ant thrushes, robin chats and small hornbills follow the ants and snap up any insects escaping from them.

Giant plants Above the montane forest is a narrow belt of bamboo, interspersed with a few of the same forest trees as grow lower down the slopes. Animals that inhabit the forests may also be found in the bamboo zone, although most of them do not eat the bamboo itself. Buffaloes, elephants and black rhinoceroses are all common, together with duikers, and the occasional bongo and forest hog. But there are no specialized bamboo-dwelling small mammals like the rats and bats that have evolved in the bamboo jungles of south-eastern Asia; and no animal that feeds almost exclusively on bamboo like the giant panda. As a habitat this belt is remarkably barren. The bamboos may be ten or more metres tall, with nothing but a few ferns and tiny herbs growing beneath them. About every thirty years they all flower simultaneously, producing quantities of seed, which may result in plagues of forest rats. After seeding the bamboos all die. When the seeds germinate the young plants form almost impenetrable thickets. Elephants browse the tops of the seedlings, but otherwise nothing is attracted to the bamboo zone, only to the forest trees within it.

On three of the Virunga volcanoes, however, and in the Kayonza Forest of Uganda, gorillas feed on bamboo at certain times of the year when young shoots are sprouting. Gorillas are entirely vegetarian, and eat a great variety of plants— lemonwood, wild celery, goosegrass, muskthistle, wild and cultivated bananas and even stinging nettles which they pick and eat, apparently without minding being stung. On the volcanoes Mgahinga, Sabinio and Muhavura, the mountain race lives mainly in the luxuriant *Hagenia* and bamboo forests around 3,000 metres. About 150 centimetres of rain a year produces thick ground cover and abundant forage and gorillas are most numerous and at home here. In the Kayonza, known as the Impenetrable Forest, rainfall is only about 100 centimetres but the constant low clouds and high humidity produce dense vegetation on the steep ridge sides, and swamps in the valleys. The gorillas, which may be of the lowland race here, move up onto the more open tops of the ridges by day, to feed and to bask in the sun if it breaks through the mist. At night they move down into the valleys to avoid the colder air of the heights.

Gorillas are not found outside dense humid forests but chimpanzees, also forest animals, can live in very dry, scrubby areas with few trees. They are not found on the Virunga volcanoes but the long-haired race is not uncommon in many forests in East Africa, especially in the Budongo Forest of Uganda. Chimpanzees are mainly vegetarians, but they sometimes eat insects and birds' eggs, and in Tanzania they stalk and kill young baboons, baby antelopes and birds. On the Ruwenzori range they live in montane forests up to 2,700 metres, where they climb trees with far greater agility than the heavier gorillas, but they also spend much time feeding and moving about on the ground.

Above the bamboos, from around 3,500 metres, occur heather forests and the giant groundsel zone. On ridge tops, or where the soil is poor or rocky, tree heathers abound. Their trunks and the ground

Giant groundsels near the summit of Mount Elgon, sending up their spikes of daisy-like flowers. These plants are unique to the East African mountains; among them grow clumps of everlasting daisies, a link with the flora of South Africa.

A natural rock garden of small red hot pokers, two species of dwarf everlasting daisies and other tiny alpines, flourishing above the giant groundsel zone among golden mosses and lichen-covered rocks.

are thickly cushioned with green, yellow and gold moss, and their branches are hung with the tattered grey beards of lichens. In the valley bottoms huge bogs are filled with moss and tall tussocks of sedge, with the occasional giant lobelia standing like an obelisk. On well-drained slopes, richer soils support small woodland trees, chiefly tree-sized St. John's worts with lovely yellow or orange flowers, and *Hagenias*. Beneath these grow brambles, wild celery, yellow alyssum and banks of everlasting flowers, pink and white and yellow and red. Here the first of the tree groundsels appear.

With increasing altitude the scene is gradually dominated by these giant groundsels. There are several species, but they all look much alike and all send up flowering spikes to about 5 metres. Sometimes they grow singly, sometimes in such profusion that they form groundsel forests. The plants have a thick trunk-like stem and great rosettes of cabbage-like leaves, each over half a metre long and silvered over with fine hairs. As the plant grows, the dead leaves remain attached to the trunk in a ruff, always dry in the centre, in which small birds roost at night. The flower spikes shoot up in a profusion of tiny daisy-like yellow flowers. Among the groundsels grow graceful torch lobelias, also with their leaves in rosettes, and with tall flowering spikes of pale blue. The massing of foliage in low dense rosettes, which is very common in alpine plants, reduces water loss. The covering of silvery hairs also holds in moisture and is a barrier against the strong ultra-violet radiation at these heights. Other species of tree St. John's wort also grow here, and there are acres of everlasting flowers over a metre high.

The extent and purity of the various vegetational zones varies from mountain to mountain. Among heather forests, those of the Ruwenzori are the most spectacular; in the dripping misty half-light, the lichen-festooned forests have a strange unearthly quality. Of bamboo forests, that on Mount Kenya is the most extensive, with purer stands. On Mounts Kenya, Kilimanjaro, and Elgon and on the Aberdares there are extensive areas of open moorland, rich in herbaceous plants such as saxifrages and a lovely scarlet-flowered wild gladiolus. Above about 4,000 metres on all the mountains, giant lobelias and groundsels start to thin out and everlastings are represented only by dwarf species covered with white woolly hair. Gradually the vegetation is reduced to mosses and lichens until just below the snow-line nothing can grow and no animals live permanently.

29

The klipspringer is at home on mountain cliffs and rocky ledges up to the snow-line. At sunrise it stands on a rocky pinnacle, poised on its points, soaking up the sun's warmth to dissipate the coldness of the alpine night.

Mole rats are often very abundant in grassy meadows at high altitudes, tunnelling just below the surface and feeding on grass roots. They are one of the most important prey animals of the area, providing food for all predators from leopards to mole snakes.

The alpine zone The climate of the heather and alpine zones is extremely harsh; it can be very hot indeed on cloudless days, while at night the temperature plummets down to give frosts before dawn. These are quickly dispelled in the first hours after sunrise, but after mid-morning clouds and mist begin to creep up the mountains and rain falls, sometimes as hail or snow. Nevertheless, many mammals are plentiful at these altitudes. Bushbuck, duikers, even eland, have been seen up to 4,000 metres. Buffaloes are abundant; they have plentiful grazing in forest glades and on the higher alpine slopes; trees to shelter among; mud for their wallows; and above all, peace, from lions and from man. At high altitudes they develop thick shaggy coats as a protection from the cold night air.

Among the high mountain crags live klipspringers, feeding on mosses and succulents. They have the compact build of the alpine chamois, and their elongated hooves, the texture of hard rubber, give them a sure-footed grip on the rocks. But the most typical animal of high mountains is the hyrax or dassie. It lives among outcrops and rocky boulders, running easily up and down smooth almost vertical surfaces, its rubbery pads kept moist by glandular secretion. If the hyraxes themselves are not visible their presence is apparent from shiny surfaces on rock faces which catch the sun like sheets of mica. These are formed by the deposits which drip down from the hyraxes' latrines. Hyraxes regularly use one or two spots for

this purpose. During rain the lavatories are flushed out, so that fertilizer is carried down to the plants growing around the base of the rocks. The hyraxes feed on these plants, thus operating a small closed fertility cycle.

There are three genera and a dozen species of hyraxes, but in the field it is almost impossible to distinguish between them, they are all so alike in size, shape, colour and habits. Two genera live in rocky habitats, while the third lives in high forest trees. Rock hyraxes are extremely sociable, living in large colonies, like other ungulate herds. They are very vocal, too, calling and answering each other in high-pitched mewing and cackling calls which bounce and echo off the rocks. Both genera of rock hyraxes are entirely diurnal, but tree hyraxes are nocturnal and not as gregarious. By day, tree hyraxes sleep very high up in hollow trees or dense foliage, becoming active as darkness falls. They, too, are extremely vocal, yelling a series of very loud, harsh, creaking calls, terminating in drawn out piercing screams.

Both tree hyraxes and the two genera of rock hyraxes co-exist in scattered and isolated populations in East Africa. Generally it can be assumed that hyraxes among rocks are rock hyraxes and hyraxes in forest are tree hyraxes. But on the Ruwenzori and Mount Kenya the tree hyrax lives among cliffs and boulders in the upper treeless zone. Not only does it here occupy the niche which rock hyraxes fill elsewhere, but it has adopted the

The rock hyrax lives among jumbled boulders where caves and crannies provide a safe refuge from leopards and Verreaux's eagles, and insulation from the extreme climate.

habits of rock hyraxes, living in large colonies and coming out to feed and bask in the daytime.

Leopards are common at high altitudes. They feed on klipspringers, hyraxes, bushbucks and duikers. They have even been recorded visiting up to 4,500 metres on Mount Kenya to catch small rodents, possibly grass-eating vlei rats or striped mice. Mole rats are often very abundant in high grassy plateaux. These are very successful burrowing rodents with enormous orange incisors, and they occur on alpine moorlands up to at least 3,500 metres. They feed on grass roots, tunnelling through the soft ground and throwing up molehills. They are quite noisy underground, digging and chewing, and tapping either with the teeth or by stamping the hind feet. Several kinds of burrowing animals produce a drumming noise with their hind

Jackson's francolin is confined to the forest and bamboo zones of Mount Kenya and neighbouring highlands. Like other francolins it feeds by raking over the leaf litter and animal droppings to find small invertebrates. Each bird must turn, aerate and mix tons of surface soil and debris in its lifetime.

In the high mountains the augur buzzard preys chiefly on mole rats, watching for them from a lookout post or while soaring overhead. At lower altitudes it takes other kinds of rats, as well as the occasional lizard or snake.

feet, but the mole rat's signals are slow tap-tap-taps. Their underground progress is also marked by the heaving up of earth above surface runs and molehills, and by grass stems disappearing underground.

Mole rats fall prey at high altitudes to augur buzzards and tawny eagles by day, and to mountain eagle owls and servals at night, as well as to the occasional leopard. Mole snakes pursue them underground. So at no time of the day or night are they free from predation, above or below the ground; and yet they continue to be very numerous.

The most conspicuous small birds in alpine moorlands are the dark brown mountain chats that flit about among boulders and perch on clumps of everlastings, flirting their tails and scolding. They nest in holes under rocks or in earth banks where elephants have mined for salt. In places francolins and white-necked ravens are locally common. The rather rare but spectacular Verreaux's eagle, a coal-black, white-rumped bird of the mountains, feeds almost exclusively on hyraxes and perches high on lofty crags.

Alpine swifts nest on steep cliff faces almost up to the snow-line, but fly down 3,000 metres every day to the foot of the mountain to feed on the more

abundant insects there. Various species of sunbirds occur at different levels, feeding on the nectar of lobelia and St. John's wort flowers, and weaving their tiny hanging nests from the wool of various alpine plants. At night they roost in the giant groundsel ruffs. The problem of heat loss during sleep must be an acute one for such tiny birds at high altitudes. Perhaps, like the humming birds of South America, they are able to avoid dying of exposure by becoming cold-blooded. At night, some humming birds fall into a torpid state during which their body temperature falls to about 5°C. They lose less heat in this state because the difference between body and air temperature is very much less than usual.

There are few other small animals at these heights. The high-casqued chameleon may be found among lichen-covered bushes, and surprisingly, frogs occur quite high up; the cold mountain streams may be full of tadpoles. Freshwater crabs are plentiful, too, which accounts for clawless otters visiting alpine bogs. There are no indigenous fishes—they have been unable to surmount the rapids—but introduced trout flourish in the mountain pools.

Different species of sunbirds occur at all altitudes from sea-level almost to the snow-line, wherever there are nectar-producing flowers. Some are confined to high altitude moorlands where they feed on giant groundsels, lobelias and red hot pokers; others live only in forest; some, such as this scarlet-chested one, are found in a variety of habitats. Their long, slender beaks are specially adapted for probing tubular flowers such as those of aloes and they are of prime importance in ensuring the cross pollination of the plants on which they feed.

This high-casqued chameleon is unusually well camouflaged among the silvery-pink flowers of everlasting daisies because it is in the process of sloughing. When not shedding its skin it is usually found among small shrubs or even on rocks where its bright green and reddish colouring blends with leaves, lichens and mosses.

The Great Rift

The Great Rift is a gash in the Earth's surface 5,600 kilometres long, running from south of the Zambezi Delta northwards through the highlands of East Africa and Ethiopia to the northern end of the Red Sea and beyond. It is part of a much greater series of submarine rifts along the floor of the Indian Ocean. Two branches of the system run through East Africa, the Great Rift and the Western Rift. At one time it was considered a very remarkable geographical feature and was presumed to be still widening, to become an ocean in another few million years. But it is now known to have widened by only about 9·6 kilometres in 20 million years, while the Red Sea widened 32 kilometres during the same period, and far faster widening has occurred in other places.

In its youth the East African rift was more than a kilometre deeper than it is today, but it has been filled in by the hundreds of volcanoes along its length. Nevertheless, some spectacular scenery remains. Vertical cliffs face each other, sometimes only a few hundred metres, sometimes sixty kilometres apart, across a flat plain dotted here and there with the rounded cones of extinct volcanoes. Sometimes it looks as if a strip of ground has simply dropped a few hundred metres into the earth. In places the cliffs run for many kilometres through otherwise flat bush country.

These cliffs are the home of rock hyraxes and klipspringers, leopards and Verreaux's eagles, like the cliffs and rocky places on the mountains. They are also the haunt of Schalow's wheatear, a very local Rift Valley bird; and of the mountain reed-buck which lives on the grassy slopes and scree at the foot of the cliffs. Baboons climb noisily up to the high cliff ledges for the night. Swifts, too, seem to roost high on these cliffs. They collect towards evening in vast wheeling, screaming mobs,

Flamingos epitomize the Rift Valley soda lakes: nowhere else in the world are these magnificent birds concentrated in such numbers nor seen in such glorious settings. In the brief golden glow that precedes the sunrise, the lesser flamingos are feeding quietly, concentrating on filtering algae from the soupy water. But the greater flamingos, coming into breeding condition, are wading about, turning their heads smartly from side to side as they work themselves up into the state of intense communal excitement which culminates in the wings-out courtship display (page 124) and mating.

tearing through the air above the cliff tops. Long charcoal-grey flight-feathers spiralling down testify to the loss of a few individuals as prey to peregrine falcons, but these rare raptors have no appreciable effect on the vast numbers of the swifts.

Vultures, particularly the griffons and the Egyptian vulture, roost and nest on cliff ledges. But the lammergeier is the true bird of great cliffs and gorges. In East Africa it is rare and very local, but a pair regularly roosts in the spectacular Hell's Gate Gorge near Lake Naivasha in Kenya. In the early morning they spiral up slowly and majestically, their long, narrow wings taking them skywards on the faintest up-draughts from the cliffs.

Soda lakes In prehistoric times, the floor of the Great Rift was filled with swamps and great lakes, but today all that remain of these are plains of white dust, deposits of diatoms, and a line of much smaller alkaline lakes. Some of these are fed by rivers, others by hot springs. The volcanic rocks of the escarpments and mountains on each side of the Rift Valley are exceptionally rich in carbonates. The water permeating through is charged with salts, chiefly sodium bicarbonate, which become concentrated in the lakes. Some have no outlets, but their evaporation rate is so high that, except in years of exceptional rainfall, they are actually shrinking. A few of the lakes are now almost dry: Lake Magadi is a flat, blinding-white waste of crystalline soda, 5 metres deep in places, overlying black mud. It is streaked here and there with pink algae, with pools of wine-red water, but from a distance it looks like an ice-rink, not a lake. It is the second largest expanse of pure washing soda in the world, and is extensively mined; but the inflow of soda from all its hot springs—over 400 tons a day— is greater than loss from commercial exploitation.

The hot springs of Magadi are the home of a remarkable dwarf cichlid, *Tilapia grahami*, which was forced to adapt to this extreme environment as the lake shrank and became more alkaline. The fish's ancestral forms grew larger and inhabited a much deeper, probably freshwater, lake; beds of their fossils have been found on escarpments above the present Lake Magadi. Today *Tilapia grahami* lives in water with a high specific gravity as well as high alkalinity and high temperature. Water temperatures in the springs vary from lukewarm to not far off boiling, but the fish are restricted to pools and lagoons with a temperature of about 38°C. Below this they become sluggish and cannot breed; only two degrees hotter and they die. They

feed by browsing blue-green algae which grow on the gravel bottom, rasping the gravel clean in water of optimum temperature up to a distinct browse-line, where the water becomes too hot for them. But the algae grow most luxuriantly in the hotter water and some fish are tempted by greener pastures to cross the browse-line, as the many corpses on the green bottom testify. Another similar tilapia species occurs in the hot springs of Lake Natron.

Lake Hannington is also fed by springs, which well up among rocks along the shore, crystal clear and only three or four degrees below boiling point. From these geysers the water streams down to the lake through channels in the rocks, cooling as it goes. Close to source the channels are encrusted with mineral deposits, but a little way down the shore they become overgrown with dense streamers of brilliant green algae, although the water is still very hot. In a nearby grassy meadow are deep pools, slightly less hot, with water deceptively still and clear, tempting animals to alight on the surface. White bones gleaming at the bottom show that birds have made the fatal mistake as have the dragonflies, water scorpions and other insects whose boiled bodies collect around the rims of the pools.

Only a few forms of animal life are able to live in the strongly alkaline waters of the Rift Valley lakes, but lakefly larvae, small water-beetles, water-boatmen, and copepod crustacea flourish in a soda solution. The murky waters are often thick with "water fleas", but are also a kind of living vegetable broth. Their high carbonate content, the warmth and sunlight, produce ideal conditions for the rapid multiplication of microscopic single-celled plants: diatoms, with beautifully sculptured siliceous skeletons, and blue-green algae, the most primitive plant occurring today. In some of the more alkaline lakes blue-green algae occur in such abundance that the water looks like pea soup —a fertile pasture for animals that can use it.

Flamingos Flamingos are the only large animals adapted for filtering suspended algae from the water. The lesser flamingo is the more highly specialized, and lives almost entirely on single-celled plants. The greater flamingo also filters algae, but in addition it takes in a lot of mud rich in organic matter, and sieves shrimps and small molluscs from the bottom. Since flamingos are the only birds making use of this food, the abundant algae are able to support them in vast numbers. Some three million flamingos live in the Great Rift,

Schalow's wheatear flits among rocks in the Great Rift. Lichens are the first living things to colonize bare rocks, and can do so because they are dual plants made up of filamentous fungi and algae. The fungus part absorbs and holds moisture, and produces an acid which disintegrates the rock, allowing the plant to anchor itself. The algae manufacture food for the whole plant.

Small rounded cones and craters of extinct volcanoes in the floor of the Rift Valley.

Sheer vertical cliffs in the Rift Valley are the home of peregrine falcons and klipspringers. Baboons climb each night to high ledges; vultures and lammergeiers nest here. Among tumbled rocks at the foot live rock hyraxes, and on the grassy scree, mountain reedbucks.

as many as occupy similar habitats throughout the rest of the world. They extract between them about 6 tons of algae per acre each year, yet the algae multiply at such a rate the water is never perceptibly less green.

The flamingos' habitat is as hot and shadeless as any desert, and flamingos need to drink as regularly as any other bird. They come once a day to the outlets from the boiling springs where the water is relatively fresh and has cooled to about 75°C. They are just able to drink from the surface of this water although they cannot paddle in it. After drinking they go down to the edge of the lake to bathe, bobbing and fluttering in the water, then retire along the shore to preen. Below the springs there is constant excitement; all day birds are flying in to drink and splash while those that have bathed move on.

Large pythons among the rushes at the edge of a lake will occasionally grab a flamingo as prey, but the nature of their environment keeps flamingos otherwise fairly free from attack except from the air. Tawny and steppe eagles sometimes take them in flight with a powerful falcon-like stoop, and where the lake has no fish, fish eagles may feed almost exclusively on flamingos. Adult fish eagles

that regularly feed on flamingos can be distinguished from fish-eating fish eagles by the colour of the feathers of throat and breast. Normally white, the feathers of flamingo-eaters are stained pink by the carotin in the flamingos' bodies.

But the chief predator of the flamingo is the marabou stork. It is primarily a scavenger, but around the soda lakes, where flamingos are not only exceedingly numerous but particularly vulnerable to the marabou's method of attack, it is a predator in its own right. A flamingo's one defence against the marabou is flight; once on the wing it is safe. But when it is paddling in deeper water take-off is not instantaneous. To become airborne a flamingo must first pull its legs up out of the water with a jump, then by half flying, half running, get enough lift to take it clear of the surface. When hunting, a marabou strides purposefully along the edge of the lake, between the flamingos and the shore, herding its prey into deeper water where take-off is most difficult. Suddenly it launches itself into the air and flaps out towards the flamingos which are now wading in a tight pack. Immediately they see this great dark shape heading towards them there is panic, and a brief confused struggle to take off. Those on the outside patter unimpeded across the open lake away from the marabou, but those on the inside of the throng must wait until the press has cleared a little before they can take off. Very often they all manage to get clear and the marabou returns to the shore to hound another section of the vast band. Finally, one bird on the inside fails to make a quick enough get-away, and the marabou pins it down in the water. Once a flamingo is wet, take-off is even more difficult, so the marabou has no need to make a swift clean kill.

Other fauna of the soda lakes The fishless soda lakes support few other birds beside flamingos. Avocets and black-winged stilts find enough invertebrates by their own peculiar fishing techniques—avocets slicing through the water with upturned bill, stilts picking at the surface of water

Left: Adult steppe and tawny eagles are difficult to distinguish in the field. Both prey on lesser flamingos around the soda lakes.

Below left: Lesser flamingos gathered to drink the fresh water by the flat, grassy meadow where Lake Hannington's boiling springs well and steam. The farther spring close to the lake shore maintains a channel through the flamingo band; the birds can drink but cannot paddle in really hot water.

Below: The flamingo's only defence from the marabou is to take off, pattering over the water; once airborne it is safe.

The tiger beetle is an active predator, running swiftly about on sandy lake shores and pouncing on any other insect, such as this caddisfly, which it can grab in its large jaws. Perfectly camouflaged, it becomes virtually invisible as soon as it stops running.

The long-pincered earwig lives in burrows under logs or beneath boulders of jagged volcanic rock on the shores of soda lakes. Like the tiger beetle it is a voracious predator, but it emerges chiefly at night, whereas the beetle is active by day.

The fish-eating birds of Lake Nakuru employ different fishing techniques.

Left: Grebes are the most perfectly aquatic of all flying birds. They pursue small fish under water, propelling themselves solely with their webbed feet. This African dabchick or little grebe is offering its mate a small *Tilapia grahami*.

Right: Cormorants also pursue fish under water, but use their wings as well as their webbed feet for swimming. They have truly amphibious vision, but the waters of Lake Nakuru are so murky with suspended blue-green algae that birds must also be able to locate fish acoustically. After fishing, the white-necked cormorant holds its wings out to dry the sodden feathers.

Right, below: Pink-backed pelicans fish on their own, plunging the bill in whenever they see a fish near the surface. If the strike is successful, the pelican withdraws its bill carefully, allowing the water to drain out and trapping the fish in its pouch. The bird then throws back its head and gulps down the fish.

Far right: The avocet fishes by slicing through the water with upturned bill, catching various invertebrates and fish fry.

or mud. Dabchicks catch water-boatmen underwater; blacksmith and spurwing plovers, little stints and other migrant waders find food—aquatic insects—at the lake edge. Around the lake, over the sedges and yellow grass, mosquitos, midges and lakeflies may dance in clouds, following game like a moving halo, and these tiny flying insects are hawked by martins and swallows.

The grey sandy beaches of the soda lakes, strewn with jagged black lava rocks, look inhospitable deserts, yet some insects have adapted to this environment. Brown-and-yellow spotted tiger beetles run swiftly across the sand, pouncing on any other insect they can overpower. Another carnivore is the large, long-pincered earwig which hides under stones by day, even close to the water's edge, and comes out at night to feed. At dusk, the piercing stridulations of mole crickets accompany the twittering squeaks of the insectivorous bats.

Lakes Elmenteita, Hannington and Nakuru have no endemic fish, but in the early 1960s *Tilapia grahami* was introduced into Nakuru from Lake Magadi as a mosquito-control measure. The fish multiplied so rapidly that their teeming descendants now attract thousands of fish-eating birds to the lake where none came before. There may be a thousand or more commuting pelicans there, a thousand resident pairs of cormorants and darters, uncountable numbers of dabchicks, herons of several species, fish eagles, scores of storks and spoonbills, hundreds of gulls and terns, and occasionally a few skimmers or an osprey. These birds all employ different fishing techniques and fish at various levels, but they are all feeding on the one species, apparently without over-fishing.

The sun filters through the canopy of a luxuriant forest of tall feverthorn trees, encouraging an undergrowth of creepers and shrubs, and picking out the conspicuous rump-pattern of a female defassa waterbuck.

Feverthorns Near many of the soda lakes are stands of beautiful, tall feverthorn trees, in some places forming luxuriant forests. Their flat-topped crowns are not dense but are covered with a fine, lacy bright-green foliage through which the sunlight filters, so that an undergrowth of shrubs and creepers flourishes in the semi-shade. Colobus monkeys, turacos and trogons feed in the canopy and the trees are full of the calls of birds; wood-hoopoes, warblers, emerald-spotted wood doves, paradise flycatchers and bell shrikes. In the early morning and evening the horizontal sun cuts through the flimsy foliage and splashes the pale yellow trunks with patches of sulphur.

Feverthorns are found also along the banks of streams and rivers throughout the bush country, but they are most luxuriant around the lakes. The belts of woodland are rather narrow and many herbivores concentrate in and around them. Some animals inhabit main vegetation zones, pure grassland or dense forest, but very many prefer the varied vegetation found where two zones meet. At the edge of feverthorn woods, where dense undergrowth gives way to grassland, animals find a greater variety of food plants. They can bask in the sun, or keep cool in the shade. The woodland animals can creep among the dense undergrowth, while the plains animals can dash away in the open. Impala, waterbuck, warthog, Kirk's dikdik, black rhinoceros, elephants, bushbuck and vervet monkeys can all be seen most frequently at the edge

Kirk's dikdik is a dainty little antelope peculiar to dense bush country. It is often seen among thick shrubbery and undergrowth at the forest edge, where it is preyed upon by leopards and the large forest eagles.

Vervet monkeys often live in feverthorn forest, though not in the canopy. They may travel far from trees to forage on the ground but return to rest and groom each other in thick creepers and bush at the forest edge.

42

of woodland, while buffaloes lie up in the forest during the day and come out into the grasslands to graze at night. The phenomenon of animals concentrating where two zones meet is known as *boundary effect*.

The northern lakes The two northernmost lakes in the East African Rift, Lake Baringo and Lake Rudolf, are less alkaline than the southern soda lakes, and have a more varied fish fauna. Baringo contains quite large tilapia, mudsuckers and big catfish. Rudolf, a very much vaster stretch of water, also contains two species of very big and powerful predatory fishes. Nile perch, the largest non-marine fish in Africa, grows to about 2 metres in length and may live to over twenty years. Big Nile perch feed on nothing but other fishes such as big

The black and white colobus monkey eats mostly leaves and lives exclusively in the canopy of tall trees in forested areas, montane as well as feverthorn forests.

cichlids, smaller Nile perches, and tigerfish, a very large sharp-toothed characin which is almost as ferocious a predator as the Nile perch itself.

In most rivers and lakes of East Africa crocodiles have been almost wiped out for the sake of their skins. However, there are two places where they still enjoy something of their former abundance: the Victoria Nile in Uganda below the Murchison Falls, where they are protected to some extent in a national park; and Lake Rudolf. In this lake they are isolated by the surrounding deserts, but more importantly, the alkalinity of the lake water causes their skins to lose much of their commercial value, so they are not worth poaching.

In most localities, crocodiles feed mainly at night, though at Rudolf they are prevented from doing so by high winds that get up three hours after dark and blow until morning. Baby crocodiles feed mainly on insects—water scorpions and other aquatic species which prey on small fish, or land insects that drop onto the surface by mistake. From insects they graduate to small fish, and then to birds and mammals that fall into the water or can be taken when they come down to drink. Medium and large sized crocodiles also take Nile perch. They are therefore of great economic as well as ecological importance, particularly when they are very small or very big for they then feed largely on the predators of commercially important fishes.

African cichlid fishes are extremely prolific and fast growing. One species in particular, the Mozambique mouth brooder, is an important food fish and has been transported all over the tropical world. Other species have been introduced locally in East Africa into waters where they are not endemic, so the geography of cichlid species is often complicated. These cichlids owe their high rate of reproduction to a remarkable breeding behaviour. The females brood their eggs and young fry in their mouths, a method which ensures high success in brood rearing. In addition, the young fishes mature early. They start spawning at two to three months old, and are capable of multiplying a thousandfold in three months. Some species are quite small, but others may be 30 centimetres or more in length, and these are very valuable food fishes.

The Western Rift Lake Tanganyika, together with Lake Malawi (Nyasa) to the south, and Lakes Kivu, Edward and Albert to the north, lie in a very deep trench in the western arm of the African Rift. Lake Kivu is the most beautiful of this chain of lakes; it has deeply-indented fiord-like bays, formed when lava flows from the Virunga volcanoes dammed the valley which once drained northwards to the Nile. Its fish fauna, however, is limited and reminiscent of its previous nilotic connections, for though the lake now drains into Lake Tanganyika—which has a truly fantastic fish fauna—impassible rapids on the connecting Ruzizi River prevent new species from reaching Lake Kivu.

Both Lake Malawi and Lake Tanganyika are enormously deep; the latter has been sounded to a depth of 600 metres. Malawi is not quite so deep, but its bottom is still below sea level. In both, the lower layers of water are lifeless, for though they are rich in dissolved minerals these are not brought up

The eggs and fry of the Mozambique mouthbrooder are given a high degree of parental care, the female brooding them in her throat pouch. When still very young the baby fish are spat out into the water for short periods, but if danger threatens they rush back into their mother's mouth.

to the surface by convection currents. However, the surface layers and lake margins support a remarkable aquatic fauna, including many endemic species such as a surface-skating, flightless caddis fly, aquatic snails, peculiar crustacea and the completely aquatic water cobra. They have also produced a great proliferation of fish species, especially among the cichlids, no doubt because of the enormous size of the lakes and their long geographical isolation. Lake Albert has only seven species of cichlid, of which two are endemic. By contrast, Lake Tanganyika has ninety species, all endemic except one, and Lake Malawi has 175 species all endemic except for four.

Golden cichlids show distinct sexual dimorphism, the male having the black belly. This species is one of 171 cichlid species endemic in Lake Malawi.

Bass and tilapia in Lake Naivasha support a profusion of fish-eating birds. This purple heron, only one of many heron species frequenting the lake, waited for three hours before the large tilapia it was watching swam near enough for a strike.

Swamps and marshes Lake Naivasha is unusual among the Great Rift lakes in being freshwater. Fifty years ago it swarmed with a small indigenous fish, a top minnow. In 1926 a species of tilapia was introduced from the Athi River, and immediately began feeding on the top minnow. Later, bass from North America were also introduced, and between them these two very much larger species have apparently wiped out the top minnows. However, the big fish now support a profusion of fish-eating birds, including several, such as the purple and goliath herons, not generally found on the alkaline lakes.

Lake Victoria, the largest lake on the continent, and equidistant between the two arms of the African Rift, has a similar and wonderfully varied bird life. Like Lake Naivasha, it is dominated by a single plant species, a giant sedge, papyrus. The straight, leafless, triangular stems rise 5 metres high, with a mop of fine bracts at the top, and spring from a mat of dead stems and living roots which may be floating. The stems grow so thick and interwoven that a man has to cut a path through them. Scarcely any other plants can grow among papyrus, and few animals live in it. Only the extraordinary whale-headed stork or shoebill inhabits the vast

One of the most spectacular fishing birds in East Africa is the African fish eagle. It watches from a tall tree near the shore, then, flying low over the water, grabs a surfacing bass in its talons.

The tiny malachite kingfisher hunts from a low lake-side perch. It repeatedly whacks the tiny tilapia on the log before swallowing it.

papyrus swamps which surround the lake and stretch along the Nile basin to that greatest of all swamps, the Sudd, in southern Sudan. The shoebill hunts frogs and fishes deep in the swamps, and breeds there. When travelling from one swamp to another it soars very high in the thermals, so in spite of its great size it is very rarely noticed.

Underneath the floating islands of papyrus the water is dead, deprived of light and oxygen. In between it is often covered with great rafts of waterlilies, with lovely blue and purple flowers that open during the morning and close again by mid-day. Among the lilies there is far more animal life; in the clear pools between the leaves tilapia can be seen browsing or basking just beneath the surface; herons wade where the water is not too deep and kingfishers perch on the papyrus stems around the edges of the pools. Ducks and geese swim here to feed and wagtails run about on the lily leaves catching small flies. Dragonflies abound, and reed frogs set up their "pinkle-pinkle" chorus at dusk. Throughout the day land birds flock to suitable drinking points; even vervet monkeys visit the pools, climbing up the tall papyrus stems until these swing over with their weight to form convenient bridges across the channels. The monkeys come in search of lily pods, which are a great delicacy. They walk out along a stout papyrus

stem, lifting the lily leaves and peering into the water, then reaching in a long arm to pick the prized pod.

The most typical bird of the waterlily-papyrus marshes, and the best adapted to this habitat, is the African jacana, called the lily-trotter or lotus bird. Its very long toes and claws enable it to walk on the floating lily leaves without sinking in. Jacanas find all their food among the lilies, lifting the leaves with the bill and holding them folded over by treading on them with one foot. They find aquatic invertebrates, insect larvae, and tiny reed frogs; they also relish the lily pods, though they are not large enough to collect these easily. Even red-knobbed coots, which can dive down for the pods, have difficulty in breaking them from their anchoring stems. The only bird that can collect a pod with ease is the purple gallinule, a much heavier bird than the jacana, which has to run across the lily leaves to avoid sinking in at each step. When a purple gallinule brings a lily pod to the surface, it snips it neatly from the stalk, then, half-flying, it runs over the lily leaves to a firmer perch. There it holds the pod in the toes of one foot while it chops it to pieces. Smaller, would-be pod-eaters—jacanas, moorhens, Allen's gallinules and black crakes—circle around hoping to pick up some edible scraps. Sometimes they are lucky and the purple

46

Islands of floating papyrus break off from the main mass and are blown across the surface of the lake. Like bamboos, papyrus is sterile as a habitat—nothing eats it and hardly anything can live among its dense tough stems.

Tiny reedfrogs abound among waterside vegetation. During the day they sit motionless for hours on a leaf or a reed, legs tucked in, exposed to the hot sun. At night they whistle in shrill chorus and feed on insects such as roosting damsel flies.

gallinule tires of the pod half-eaten. It seems deliberately to vary its diet, eating some pod then feeding on pink persicaria flowers or aquatic weeds before searching again for a pod. Thus all the pod-eaters get some occasionally, though none lives wholly on pods.

Papyrus swamps give way in places to reed and bullrush marshes. Like the waterlily pools, these are full of birds: snipes, crakes, night herons, coucals, ibises, egrets, crowned cranes. Some weavers commonly build their spherical nests high up on tall reeds, and the thicket rat may later take over the old nests. Marsh mongooses hunt in family packs, crashing among the reeds as they pounce on frogs or birds. And in certain swamps sitatunga are common. Their greatly elongated hooves, which splay out under their weight, allow them to walk on sodden vegetation without sinking through. They

Where two vegetational zones meet, animal life is often particularly abundant. The purple heron does not live among the papyrus, but fishes at its edge. Other marsh birds, frogs, fish and aquatic insects congregate where papyrus meets waterlilies, where shade and cover are close to open spaces with plentiful food resources.

The extraordinary long toes and straight claws of the African jacana allow it to walk easily over floating vegetation, especially waterlily leaves. This bird has found a tiny frog, possibly a waterlily frog, which lives under the overlapping leaves, glueing them together with its spawn.

move about the swamps in tunnels through the reeds, often up to the withers in water, walking on submerged vegetation. Their passage is marked by swaying bullrush heads or the loud squelching sucking noise of feet in and out of mud; but the animals themselves can remain completely hidden.

Surprisingly, in spite of such large areas of swamplands, which in former times were even more extensive, there are no native aquatic rodents in Africa. The shaggy rat lives in wet places, but not in true swamps, and so does the long-footed swamp rat. The cane rat, a relative of the porcupine, lives among elephant grass and other tall, rank vegetation in wet places, but although it can dive and swim if need be it is not truly aquatic. In some parts of East Africa, however, the coypu has become established. It is a large, South American, truly aquatic rodent with webbed hind feet, and was at one time farmed for its fur, known in the trade as *nutria*. When nutria went out of fashion many coypu were liberated in different parts of the world. In some East African lakes and swamps they flourished, feeding and giving birth to their precocious young on the floating mats of vegetation. Baby coypu may be taken by predators such as eagles, pythons and marsh mongooses, but an

adult is a stout animal with very large orange incisors, and can in any case escape simply by slipping into the water. It will be interesting to see whether feral coypu will be successful in East Africa, and eventually take over a niche that, surprisingly, is not already occupied. Or it may be that the niche in fact does not exist, since no mammal has evolved to fill it.

Hippopotamuses A variety of amphibious animals is found throughout East Africa. Around Lake Victoria and other Uganda lakes the spotted-necked otter is common; in the rivers, streams and lakes of Kenya and Tanzania the larger clawless otter takes its place. In mountain streams of the Ruwenzori range and in the Kalinzu Forest in Uganda lives the solitary, nocturnal otter-shrew, hunting crustacea, otter-like, in small, fast-flowing streams. But in every suitable lake and river throughout the region, hippopotamuses are common.

Truly amphibious, hippos can swim at surprising speed, both at the surface and moving along the bottom, and they do so with extraordinary grace. They spend the day wallowing in the water, surfacing to breathe, with much snorting and ear-

By day hippos herd together in the water to keep cool and avoid sunburn. They have a profound influence on the ecology of rivers and lakes, since the amount of grass they consume on land by night is enormous, and much of this, converted into fertilizer, is released in the water during the day.

flicking. At night they come ashore to feed, half swimming, half wading through any floating mats of vegetation, breaking them up and maintaining channels which can be used by many smaller animals, such as ducks and geese, otters and marsh mongooses. They feed on the short grasses of water meadows, cropping the sward with the effect of lawnmowers. The males maintain territories ashore, with well-worn paths and landmarks beside which the owner of the territory sprays faeces and urine. At dawn the hippos trip down to the water again, daintily, in spite of their great bulk. During the day, they deposit more dung, in the water. On land the dung has social significance for the hippos, but since it is concentrated in a few places it is not very useful as a fertilizer. In the water it is of no importance to the hippos, but is a great environment enricher. Large herbivorous fishes may nibble some of the grass fibres, but more important, the nitrogen-rich dung fertilizes the green algae on which many kinds of fishes feed.

Riverine forest on the banks of the big permanent Mara river. Such forest is particularly rich in bird and mammal species, from monkeys and turacos to the very rare Pel's fishing owl, which used this fig tree as a daytime roost.

Hartlaub's turaco is the common turaco of highland and riverine forests. It feeds on fruit, running monkey-like along the branches. Though often unseen it draws attention by its raucous calls.

Riverine Forests The banks of rivers are very often bordered by strips of riverine forest, dense evergreen trees hung with creepers, through which elephants and hippopotamuses maintain broad paths on their way to and from the water. The very rare Pel's fishing owl roosts in a big old tree by day; at night it hunts fish and frogs in shallow reaches and backwaters. Wild fig trees grow in these forests. Their small latex-bearing green fruits sprout in bunches out of the trunks and branches, and ripen all at the same time to a delicious-looking reddish-orange. Within a few days the figs have been stripped by every fruit-eater within miles: monkeys and turacos, flocks of green fruit pigeons, hornbills, bulbuls, mousebirds and many others by day. At night fruit bats chew the figs, swallowing only the juices and spitting out pellets of dry fibres which patter on the forest floor like hail.

Bush pigs and red duikers are often common in riverine forest, and troops of monkeys may travel miles along forested water courses. Sun squirrels and bush squirrels, both arboreal, also live here. Like the narrow strips of feverthorn woodlands, this is a habitat particularly rich in species, for it contains not only forest-dwellers but also species that prefer the variety of vegetation found here.

Arboreal squirrels are far less common in East Africa than are the wholly terrestrial ground squirrels, but in riverine forest the bush squirrel sometimes occurs. It is preyed upon by forest eagles so any bird of prey passing overhead sets off its churring alarm call followed by a rapid dash to safety in a hollow tree.

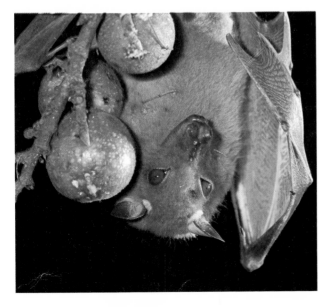

Wild fig trees in riverine forest may ripen during the dry season when other food is scarce, and provide a brief glut of fruit for monkeys and fruit-eating birds by day, and fruit bats by night.

The Bush Wilderness

The bushlands of East Africa are wildernesses of thorny thickets. They cover thousands of square kilometres of eastern Kenya and Tanzania, from the narrow coastal strip to the mountains. Here rainfall is often less than 50 centimetres a year, so the vegetation is adapted to near-desert conditions, and either taps water deep in the soil with very deeply-probing roots, or stores it in underground bulbs or tubers, swollen stems or succulent fleshy leaves. To conserve moisture, trees shed their leaves after the rains, and during the long dry seasons the bush looks dead. Grey and ochre are the predominant colours: bare twigs are grey; branches are festooned with grey-green leafless creeper stems; the dry earth is red; the scaly bark of trees is red or yellowish. In the dry season the bush is a hot and waterless place. Small river beds run only with dry sand; large permanent watercourses are few. In the rains, torrential downpours bring out a flush of ephemeral spring-like greenery and the rivers roar briefly. But the verdure and humidity soon give way again to bare twigs and drought. Yet this inhospitable region is the home of many big animals, well adjusted to the arid environment and at home in the wilderness.

The browsers The larger inhabitants of the bush are mainly browsers, for grass is often non-existent. Elephants, black rhinoceroses, giraffes, greater and lesser kudus, gerenuk, Grant's gazelles and dikdiks are the typical mammals of the bush. They are able to make the best use of this habitat because they are either nomadic, travelling long distances to water, or they can live entirely without drinking. They do not compete with each other, achieving ecological separation in their choice of terrain and by making use of the vegetation at all levels, from the ground up to about 6 metres. Giraffes are the tallest of all and the only animals that can feed on high foliage without having to push the tree over first. They may browse the tops of bushes so heavily that they make them stunted and flat-topped, while on the underside of taller trees they leave a browse line at 5 to 6 metres. In other places they keep small acacias and desert dates pruned into oval shapes. If the growth manages to get away above the giraffes' reach, however, the tree is clipped into the shape of an hour-glass.

Giraffes prefer flat country and browse chiefly on acacias. So do gerenuk. These exceptionally elegant antelopes, with their very long neck and limbs, can browse up to a height of 2·5 metres by standing upright on their hind legs with their forefeet among the branches. Kudus prefer steep hillsides and rough gullies and browse a wide variety of bushes and plants, including some generally regarded as poisonous. Tiny dikdiks browse the lowest levels of bush in all sorts of country, hilly or flat. And in all sorts of country elephants are equally at home. They need water regularly and in large amounts, so when there is nothing to drink and little but twigs to eat they seem out of place. But they are able to walk fast over long distances to get to water, and they find plenty of food by pushing over trees and stripping the branches. They are extremely catholic and adaptable in their choice of foods. So too are black rhinoceroses, which eat almost all the herbs and shrubs available in a habitat, even bitter latex-bearing euphorbias and plants, such as the thorn apple, that are highly toxic to other herbivores. They are very noisy feeders and on a quiet morning the sound of cracking branches and twigs being champed can be heard for several hundred metres.

Thorntrees Almost all the vegetation in the bush is thorny or prickly. Some acacia trees when young have particularly wicked eight-centimetre thorns; others combine long, straight, piercing thorns with shorter, curved thorns for holding. Some animals that feed on acacias are, therefore, adapted so that thorns are no problem. Many of the browsing antelopes have particularly small delicate muzzles that enable them to nibble leaves from among the thorns without getting spiked. Giraffes have long pointed muzzles and long, prehensile tongues for gathering bunches of leaves from between the thorns. Rhinoceroses eat the twigs, thorns and all, but do not look as if they enjoy them.

Some acacias have swollen thorns that look like the horns of Watutsi cattle, while others have gall-like swellings at the bases of the thorns. The best-known of the galled acacias is the whistling thorn, whose paired thorns sprout all along the twigs and branches from large round galls. The swellings are not true galls, which are growths produced by

The greater kudu prefers boulder-strewn hillsides and rough gullies. A bull, with his twisted horns, throat fringe and majestic carriage is one of the most magnificent of antelopes.

insects or mites. The whistling thorn normally bears these swellings, which are soft and at first have leaves. Later, as the galls harden and darken, the leaves disappear. Each hard old gall has a tiny round hole in it. The wind rushing over thousands of these tiny holes, amplified by the little sounding boxes of the hollow galls, produces an eerie moaning noise, which, though it is hardly a whistling, is the origin of the thorntree's name. Passing in and out through the tiny holes are small black ants which live in the galls, in a symbiotic relationship with the acacia. The ants gain some protection from the thorns, and in turn swarm onto and bite the muzzle of any animal browsing their tree. Impala are quickly moved on by the ants, but giraffe are as undeterred by ants as they are by thorns. Baboons also, are put off by neither, and will pick the galls and munch them up for the insect food they contain, spitting out the hard vegetable parts. The Abyssinian scimitar-bill also feeds on the ants, and is peculiarly well-adapted to do so. It is the smallest of the wood hoopoes, with a slender, finely curved, orange bill. It probes crannies in bark and among acacia flowers and will systematically work along the branches of a whistling thorn, extracting the ants, their eggs, larvae and pupae from the tiny holes in the galls.

Big old whistling thorntrees are also the home of acacia rats, the only truly arboreal rat of Africa. They build very large and conspicuous drey-like nests in the tops of trees, made of living twigs which they gnaw off for themselves. Berries, seeds and thorntree gum are their foods; they seem not to feed on the young galls or the ants. Acacia rats, and in other parts of the bush, tiny dormice, take the place of tree squirrels, which are not found away from forest trees. But ground squirrels, which almost never climb in trees, are very characteristic animals of the bush, running about in arid sandy areas with a curious jumping gait, their bushy tails undulating behind them. Patas monkeys, the true monkeys of the bush, have also taken to running on the ground. They are gaunt, brick-red animals, built like greyhounds and they can run very fast. They climb well, doing so to find fruits and seeds and to keep a lookout. But the usually small, almost always thorny and contorted trees of the bush are not suitable for monkeys to travel through at speed.

Larger, often isolated, thorntrees frequently have pairs of beautiful little yellow-winged bats hanging up to roost under the canopy. The bats are small and slate-grey, with enormous ears. Sometimes they fly before sundown; against the sun, their

The giraffe's long, prehensile tongue gathers and strips bunches of acacia leaves and soft young spines. Its pointed muzzle enables it to feed among the thorns without being pricked.

A small-spotted genet climbs in an elephant-damaged feverthorn tree. Genets are mainly nocturnal and carnivorous. Occasionally they will eat fruit, but generally they feed on any small animal—beetle, spider, lizard, rat, bird—which they can catch in the trees or on the ground.

The whistling thorn is the best known of the galled acacias. The swellings, not true galls, are inhabited by tiny ants which try to protect the tree from browsers and in return themselves gain some protection from the thorns.

The Senegal bushbaby may live in thornbushes, but for obvious reasons prefers broad-leaved trees. It is exclusively nocturnal and bounds from branch to branch with prodigious leaps, feeding on insects, fruits and gum.

The claw-like hind feet of the acacia rat, like those of other small arboreal animals, are reversible, enabling it to climb easily down branches head first.

Many bush-country birds perch conspicuously, on the lookout for insects or small reptiles. The aggressive fiscal shrike usually catches insects on the ground and returns to its lookout post— here the spire of a tree euphorbia—to demolish its prey.

Ground squirrels are very characteristic animals of the bush, and are often surprisingly bold, in spite of being prey to eagles, snakes, cats and other bush predators. When feeding they tripod on hind legs and tail to gain a wider view of their surroundings and guard against the danger of surprise attack.

ears and wings glow orange. At night bushbabies spring from branch to branch, feeding on insects, flowers or fruits, and genets climb to hunt roosting birds, lizards, cicadas, crickets or whatever small prey they can find.

Birds of the bush Ground squirrels are often surprisingly bold and tame. They and dikdiks sometimes seem to be the only mammals about in the bush during the day. Patas are extremely shy and many of the herbivores either feed mainly at night or are very elusive. But if the mammals are mostly retiring, birds are often noisily conspicuous. Hornbills, barbets, bush shrikes, doves, go-away-birds, weavers, guineafowl, glossy starlings can all be located by their calls, so characteristic of the East African bush. Others, such as the many colourful bee-eaters, lilac-breasted rollers, dry-country kingfishers, fiscal shrikes and the shrike-sized pygmy falcon, perch conspicuously on bushes or dead trees, watching the ground for grasshoppers and other insects, small lizards or tiny snakes.

Insects Many of the insects of the bush are beautifully camouflaged. The same species of small grasshopper may be red on red earth, grey on tufa-covered rocks near hot springs, mottled on stony ground. A very quaint, long-headed, green or buff-coloured grasshopper is found where grass is permanent. Among thorntrees, small cicadas are pale green to match the leaves, large ones mottled grey to match the lichen-covered bark. Various insects mimic acacia thorns: a moth at rest furls its forewings to resemble a straight thorn, while the hind wings are folded down to look like the swelling at the thorn's base. The very large pale green caterpillars of a saturnid moth are ornamented all over with beautiful small silver spikes that mimic thorns and also resemble the acacia's leaves.

Termites Termites are almost everywhere and are as numerous as leaves in the bush. They occur wherever there are trees, for they feed on dead wood. In the bush they consume all the acacias which have been knocked over and killed by elephants. They can eat wood because they have in their intestines microscopic protozoa which break down cellulose, turning it into sugar. Without these protozoa, termites would starve to death because they cannot digest wood on their own.

Many kinds of termites build large mounds, tall red towers which may reach a height of over 4 metres, and are a very striking feature of the bush.

The tower walls are plaster-hard and solid, made of soil and small pieces of grass, bound together by special dry, hard termite excrement. Many are thought to be of great age. The main termite nests are in the ground beneath the towers. In the cavities and galleries of big termitaria there may be hundreds of thousands of termites, mostly workers and white nymphs in various stages of growth. The soldiers are quite amazing little automata with big shiny orange-brown heads. When alarmed they vibrate their abdomens from side to side against the tunnel walls, producing a tiny rattling noise, which is instantly taken up by other termites, the sound spreading out in a subterranean hiss.

Each termite nest also contains at least one pair of reproductives. The male is normal in size, but the queen has a hugely enlarged abdomen, very

The carmine bee-eater hawks a variety of passing insects—butterflies, small grasshoppers and beetles as well as bees—on short flights from its perch.

The larger forest hornbills feed mainly on fruit, but small, dry-country species such as this red-billed one, take mainly animal food. Between them the many avian insect-eaters must consume vast numbers of insects of all sizes. In dry country it often seems remarkable that there are enough insects to support them all.

57

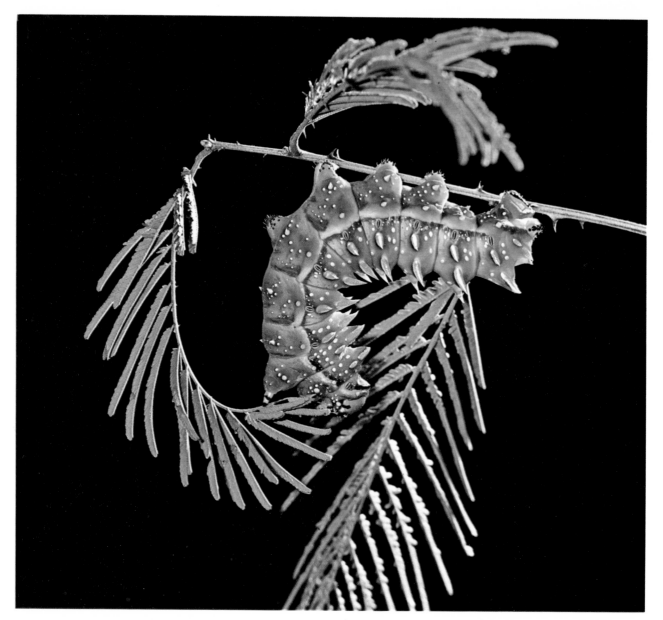

In an attempt to escape predation from the insectivorous birds, some insects are superbly camouflaged. In a world dominated by spiky bushes, many insects achieve concealment by mimicking thorns. This caterpillar of a large saturnid moth, *Gynasa maja*, feeds on acacia.

soft and delicate to the touch and in constant rhythmic peristaltic movement. She is a huge egg-producing machine, laying eight to ten thousand eggs a day at the rate of one every few seconds. The eggs are tended by the workers, which also feed and clean the queen and caress her with their palps. The workers feed and care for the developing nymphs, and feed the soldiers, which, with their enormous jaws, are incapable of feeding themselves.

With such a colossal rate of reproduction, termites are able to serve as food for a number of other animals without any diminution in their own populations. At certain times of the year, when the winged reproductives leave the nests in vast swarms, all kinds of animals gorge themselves on the brief abundance. Woodpeckers, wrynecks and other birds take a fair number of workers every day when

Above right: Termite towers are a conspicuous part of the African landscape, and the millions of termites that built them are of immense importance in the ecology of all lower lying tropical regions, wherever there are trees. Many animals benefit directly from the works of termites: dwarf mongooses, for example, use the towers as lookouts and refuges and the underground galleries as subways.

Centre right: Termites are most active above ground when the atmosphere is humid. After rain the workers get busy repairing and extending existing towers and constructing new ones.

Right: Termites are not normally active in the open by day, but a light drizzle has enabled this work-party to forage without fear of desiccation. The workers drag dry grass and petals back to the nest while soldiers stand guard.

they are poking about in dead trees for beetle grubs and other insects. Pangolins take termites, although ants are their main food. Large predatory ants known as stink ants, hunt termites individually. The smaller, equally fierce, ponerine ants go out hunting in an organized manner, in columns of up to 300; they return from their raids still in column, each ant carrying a termite in its jaws.

Termites are normally only active under cover by day, inside wood or underground, or in the little arcades of cemented earth they construct along the ground, up the outsides of trees or covering dead twigs. Here they can move about in the necessary conditions of darkness, humidity and coolness. But when the air is especially humid—immediately after rain or during a gentle shower—they are able to come out into the open to work either at rebuilding their towers or gathering food. After dark they can forage openly above ground, and are then very accessible to nocturnal small mammals such as shrews, hedgehogs and insect-eating rodents, whose droppings are often packed with the characteristic jaws of the worker termites.

Elephant shrews, though mainly diurnal, live chiefly on termites. They have a phenomenally acute sense of smell. Almost the instant their mobile snout passes over a termite, the concertina tongue whips out and in, and the insect has gone. Termites are lapped up like this at a great rate, but are not swallowed immediately; they are stored in the elephant shrew's cheek pouches, to be chewed up thoroughly later.

Termites also support two much larger mammals, both almost exclusively termite-eaters. The aardwolf, the shyest and most elusive of the two, looks like a small and elegant striped hyena, but its feeble jaws and teeth make it incapable of chewing meat. It also lacks powerful digging feet, so its feeding is limited to picking termites from the surface or out of soft soil or rotten wood. Nevertheless, its tacky tongue sweeps them up with such efficiency that it is able to consume some 40,000 in three hours of feeding.

A much more powerful animal is the aardvark or ant bear. It has strong digging claws and is a very active burrower, digging into the ground with amazing speed. Its burrows may be a complicated maze of galleries with twenty or thirty entrances, spread over a large area; or a simple shaft with only one opening. The ant bear digs out termite mounds and laps up the insects with its very long sticky tongue. In a night it may travel as much as 15 kilometres, going from one termite mound to an-

Two topi calves lay resting together on an old grass-grown termite mound while their mothers grazed nearby. When a cow joined them her calf stood up to suckle. From this vantage the mother can survey the surrounding countryside and watch for any approaching predator.

Below right: Spiny mice are gregarious and live among rocks. They feed on insects and succulent leaves, such as the fleshy *Notonia gregoriae* with its orange shaving-brush flowers. Their spiny coats may afford them some protection from snakes.

Hedgehogs are almost entirely nocturnal. Small, slow-moving terrestrial and nocturnal insects such as termites form a large part of their diet. In spite of their prickles hedgehogs are taken occasionally by Verreaux's eagle, owls and other predators.

other, and it rarely visits the same one on successive nights. Great holes around the bases of termitaria are characteristic of its work, and these holes, together with its abandoned burrows, are used as hideouts and nursery dens by a number of carnivores: spotted and striped hyenas, aardwolves, jackals and bat-eared foxes, and warthogs. Newborn warthog piglets are very sensitive to falls in temperature, and need a stable environment until they are a few weeks old. When the warthog parents have enlarged them and lined them with grass, ant bear burrows are ideal nurseries. Steinbuck too, may hide down ant bear burrows. Steinbuck are unusual small antelopes: when they are alarmed, they will often sink down into the grass and freeze until danger is past; or they may creep away and hide instead of running in the open. They

may also use ant bear warrens as nurseries and give birth underground. Even bats may hang up in them to roost, while ground squirrels, mongooses, porcupines, all sorts of small mammals, occasionally use them to hide or sleep in.

Almost as well tenanted as the big ant bear burrows are old termite mounds. These provide smaller gauge burrows for animals that find the baked earth of the bush too hard to dig in for themselves. D'Arnaud's barbets, which can dig in soft earth, sometimes nest in the ready-made hole provided by a termite ventilation shaft. Plated lizards, snakes, ground squirrels, dwarf mongooses, rats and gerbils are small enough to use the larger termite tunnels as bolt holes. Packs of mongooses may travel hundreds of metres in the safety of the subterranean galleries, and both the dwarf and banded species use old termite mounds as sleeping quarters at night, and by day as combined basking areas and lookout towers. Big antelopes too, stand sentinel on old or disused termite mounds, converted by such use into flat-topped grassy hillocks. Nile monitors lay their eggs in termite mounds after tearing open a rain-softened wall: the termites repair the breach, sealing the eggs into a predator-proof incubator. Even elephants make use of termite mounds, as scratching posts. A favourite tower may be worn smooth and rounded by their repeated rubbing and scratching.

Insulbergs In some of the most rugged parts of East Africa there occur huge isolated granite boulders known as insulbergs. These rise up out of the bush from plinths of jumbled rocks, their sides sheer and smooth, bare of any vegetation. Insulbergs are found in some of the oldest parts of the Earth, in parts of Canada and Australia as well as in Africa.

The jumble of rocks around the bases of insulbergs are inhabited by the same animals as are found in rocky places elsewhere, such as rock hyraxes, klipspringers and agama lizards. In addition, the rocks may also shelter colonies of spiny mice. Among grey rocks, spiny mice are grey; elsewhere they are usually sandy coloured. Their backs are covered with a dense coat of short, sharp prickles, perhaps a defence against snakes. The mice eat seeds and insects and the leaves of succulents, and like hyraxes, bask out on the rocks in heaps in the early morning sun.

One bird in particular is found in association with insulbergs. The freckled nightjar is a rather large, dark-plumaged species which perches by day among the rocks, well concealed by its superb camouflage. It nests directly on a flat rock, in spite of the great heat of the surface under the sun (page 84). When incubating it sits so tight it is possible almost to tread on it before it flies up. With its eyes half closed, it looks like a carcass of a bird baked dry on the hot rock.

Oases Throughout most of the wilderness water is very scarce, except immediately after rain. But in a few places water is forced to the surface by impervious underlying rock, and bubbles up in springs. There are many such springs throughout the East African bush. Some are boiling, like the springs at Lake Hannington; some may be only slightly alkaline and form pools of warm, translucent, bluish water, oases in the middle of desert bush. One of the largest is at Mzima in the Tsavo National Park. It is crystal clear and the outflow is so great that it is the main source of the permanent Tsavo River. Other springs with lesser flows soon lose themselves again in the sand. Some are large enough for fish, and even hippopotamuses and crocodiles, to live in; the smaller ones have a rich fauna of tadpoles, water-boatmen, whirligig beetles and dragonflies.

In many places in otherwise quite arid country beautiful springs well up among the rocks, oases for thirsty animals. The water is often hot and slightly alkaline—hence the blueness. Milkweed plants, on which monarch butterflies lay their eggs, have gained a root-hold in rock crevices here.

The Plains

The transition from bush to plains is often abrupt, thickets opening up and giving way to rolling grassland within a kilometre or so. In a few places in East Africa the grasslands still remain in their original pristine state, vast green or golden plateaux full of great herds of grazing game and their attendant carnivores.

Grass is the most obvious plant of the plains. The dominant species in many areas is red oat grass: when it flowers it turns the plains into a russet sea. Another sweet and nutritious species is star-grass; *Pennisetum*, on the other hand, is generally tough and unpalatable. On alkaline flats a soda-tolerant grass flourishes, while on black cotton clay, which readily waterlogs after rain and cracks when dry, there are other nutritious species. In addition, there may be many small clovers and other legumes, as well as all sorts of small herbs which are sought out by the different animals.

The smaller plains game The fauna of the plains is predominantly grass-eating. The great herds of ungulates are the most spectacular, but there are very many smaller grass-eaters, too. Among these some of the smallest, but of immense ecological importance, are the harvester ants. They are large reddish-black ants which live in underground colonies of several thousands of individuals. Each year a colony collects millions of grass-blades and grass seeds, yet the ants do not normally have any adverse effect on the vegetation. In the dry season a circular area around a nest may be quite bare of plants except for the midden of husks piled up. Yet with the first shower of the rainy season, grasses and annual seedlings sprout on this bare patch. The ants are of considerable value in aerating, mixing and fertilizing the soil in places where earthworms are few or non-existent.

As ubiquitous as grass are grasshoppers. Wherever grass grows, from sea level to the snow-line, there are probably grasshoppers to eat it. They steadily consume their surroundings with mouth-parts beautifully adapted for grinding up tough grass-blades. They are the second stage of a great food pyramid, for almost all insectivorous animals catch grasshoppers when they can, and some birds live almost wholly on them. However, grasshoppers are virtually invisible among grass. When feeding on moist greenery they are coloured bright green and when the grass bleaches they adapt their colour

Grass is the most obvious plant of the plains. There are very many species, some tough, some sweet and nutritious, but all utilized at different stages of growth by different grazers. The young reedbuck is feeding on the leaves of flowering Rhodes grass.

A young Thomson's gazelle buck, with its fine pointed muzzle, can delicately pluck small pods. Many nutritious herbs flourish among the grasses in wild pasture and supplement the grazer's diet.

A harvester ant takes a piece of grass down to its underground nest. Every year a colony of these ants collects millions of grass-blades and seeds, yet does not damage natural grassland.

A grasshopper of lush highland pasture chops through a grass-blade, then feeds it into its mouth with its forelegs. Among green grass, grasshoppers are green grass-coloured.

A superb glossy starling with a small grasshopper among flowering aloes. Grasshoppers are an abundant food and form the second stage in one of the great food pyramids. Almost all insectivorous animals catch grasshoppers whenever they can.

The long-crested hawk eagle perches in a small dead feverthorn, surveying the grass below. It lives almost exclusively on rats, which it pounces on from its perch.

Africa is exceptionally rich in birds of prey of all sizes. The bateleur eagle is one of the most remarkable. It spends nearly all day on the wing and covers two or three hundred miles, scanning the ground for birds, rats, snakes and carrion. Two hours before dusk it perches again, its colours brilliant in the low sunlight.

individually to white or yellow among dry grass or to brown, orange or black to match the soil. Some grasshopper-eaters, therefore, use other animals to find the grasshoppers for them: cattle egrets and wattled starlings follow large herbivores, and carmine bee-eaters ride on kori bustards or Abdim's storks, swooping off to snap up the flying insects disturbed by their feet. Many grasshoppers flash brilliant scarlet, yellow or blue underwings in flight, only to become invisible again the moment they settle.

Rodents are often very common in grassland, and there are several species that live on grass. The African hare and the springhare are the largest, and in north-western Uganda the Bunyoro rabbit is plentiful. There are many rat species. Grass rats live among grass almost anywhere, in very arid country or really swampy grassland, from sea level to 2,100 metres. They live in colonies and can be seen out during the day, scampering along runways in the grass, occasionally standing up on their hind legs like a ground squirrel, to get a better view. The grass rat can subsist on grass, but needs a supplementary diet of seeds to breed. The vlei rat lives entirely on grass; it has a special large caecum like a rabbit or hare, and vole-like laminated molars. With its very broad incisors it can chop down tall *Pennisetum* stems and split them to get at the soft vascular part inside. The runs it makes are most distinctive, wide and well-trodden and swept clean as the rat tramps along them on short legs, its long fur brushing the ground.

Rodents are the mainstay of the small mammal and bird predators, as grasshoppers are the mainstay of the insect-eaters. All sorts of medium-sized

The vlei rat is *the* grass-eating rat. It chops down tough stems and slits them for the soft inner parts, leaving characteristic little heaps of shredded grass. Here it is feeding on red oat grass, the flowering heads arching above it.

Bat-eared foxes setting out across the plain for their evening's hunting. Small to medium-sized predators such as these are opportunists, taking rats whenever they can but snapping up other small game such as grasshoppers, lizards, young birds or whatever they chance across.

The secretary bird, an aberrant terrestrial eagle, stalks about the plains searching out any creeping thing for prey. It roosts on the flat crown of a thorntree, waking and stretching as the sun rises.

to large birds of prey take rats and hares, and two smaller raptors, the black-shouldered kite and the long-crested hawk-eagle, live almost exclusively on rats. Mammalian predators—jackals, bat-eared foxes, wild cats, genets, even the tiny, black and white, weasel-like zorilla—take rats whenever they can. Some predators are diurnal, some nocturnal. Between them they feed on almost any creature that moves at any time on the plains, from small antelopes down to winged termites. The secretary bird, for instance, preys by day on all creeping things in the grass: frogs, locusts, small tortoises, game birds, lizards, snakes, as well as rats. When it spots something in the grass, its crest of long feathers is erected, breaking up the outline of its head and looking in silhouette from below like grass heads moving with the breeze. The puff adder, too, takes rats. It feeds mainly at night. Motionless, it waits on a run for a rat to come along, then lashes out with a lightning strike. The rat dies within a minute.

The big game of the plains Grazing insects and other small animals are almost as abundant as the grass itself, but from sheer size the big ungulates are the dominant grass-eaters of the plains. In parts of East Africa there still occur spectacular concentrations of big animals such as no longer exist anywhere else in the world. The most famous of the concentrations is found within the Serengeti National Park in north-western Tanzania and in the adjoining Mara Game Reserve in Kenya. Here over a million head of game live in 2,000 square kilometres of undulating plains interspersed with acacia woodland and riverine forest. There are many thousands of wildebeests, zebras, topis and gazelles, but also numerous giraffes, black rhinoceroses, elands and other antelopes large and small of over twenty species. Even bigger herds are found in the high rainfall grasslands of Uganda, in the Murchison Falls and the Queen Elizabeth National Parks. On both sides of the Nile live huge herds of big game, elephants, buffaloes and hippopotamuses, and enormous numbers of medium-sized antelopes. There are fewer species here than in the drier grasslands of Kenya and Tanzania, but the heavy-weights occur in such numbers that their biomass—the weight of living animals to a given area—is probably greater than anywhere else in the world, and twice that of the Serengeti.

The grasslands of East Africa are able to support this great variety and profusion of large herbivores because almost all the species are separated ecologically, some by their food requirements, some by their preferences for different types of grassland—swampy or dry, wooded or open. Waterbuck and reedbuck are found mainly in swampy grassland, and feed chiefly on grass. Impala and Grant's gazelles eat grass, but as they also browse the leaves of bushes, they are found in wooded grassland or bush.

Zebras, wildebeests, topis and Thomson's gazelles are grazers, so are mainly animals of the open plains. These four grazing species may feed off the same grass plant, but as each uses it at a different stage in the plant's growth, they do not compete. First zebras eat the outer part of the stem, which is too tough and unnutritious for antelopes but acceptable to zebras: they have incisors in both jaws for cutting through wiry stems and a gut organized to cope with a high throughput rate of low quality feed. Next topis, with their pointed muzzles, can get at the lower parts of the stems, while wildebeests' square-ended muzzles enable them to pick the horizontal leaves. Several days after this clipping the grass sprouts from the base, and Thomson's gazelles, with their small muzzles, nibble the new growth.

Buffaloes are the dominant large grazing animal in many areas. They feed mainly at night; before dawn they begin to move towards their daytime resting place, grazing as they go. During the day they lie up, usually in thick cover, then emerge again at dusk to begin feeding.

Buffalo grazing maintains a pasture in good condition. Buffaloes move along in a fairly compact group, munching mouthfuls of grass as they go, selecting not for species, but for stages of growth. They nose beneath the tougher top growth and bite off the lusher green shoots underneath. The tough, uneaten stems are trampled and act as a mulch. Next time the herd grazes that way it will find new growth stimulated by the previous cropping and easier to reach because of the trampling. Each time the pasture is used—providing it is not over-used—the grazing is improved. Patches of long grass which the buffaloes miss, either because it is unpalatable or has snags of fallen branches among it, serve to hide smaller herbivores; and in times of drought act as a useful reserve of standing hay.

Topi are also beneficial grazers because they can eat the drier grass stalks. Elephants and hippopotamuses are beneficial when they are not too numerous for they can reduce extensive areas of long, tough grasses to a short-grass sward which other animals can utilize. On a smaller scale, colonies of vlei rats improve a pasture by chopping down old stems of tussock grass. Cropping the grass stimulates its growth, so that full, even heavy grazing by a variety of herbivores keeps a pasture in an early, therefore succulent and nutritious, stage of growth. The grass is able to survive heavy mowing because its growth point is at the base not at the tip as in most other leaves.

The effects of normal grazing are subtle and often difficult to see. Only when over-grazing occurs, as when hippopotamuses become too numerous, are the effects obvious. An adult hippo eats 180 kilos of grass each night, so when at the beginning of the century, the numbers of hippos in Uganda increased enormously, the effect on the vegetation was drastic. Areas that were once grassland were grazed bare. The loss of grass prevented the spread of fires which keep down thornbush scrub, so this was able to

The puff adder's beautiful pattern camouflages it as efficiently among grass as among rocks. It does not pursue its prey but waits for a rat to approach within range. After striking, the adder tracks the rat. When it finds it, it tests it all over, then swallows it whole, head first. This adder took ten minutes to swallow the rat, and was able to subsist on it without another meal for six months.

A buffalo feeding at dawn on tall grasses at the edge of a swamp. Grazing by buffalo converts areas of tall unpalatable grass to a succulent greensward which smaller herbivores can utilize.

Burchell's zebras, wildebeest and topi graze together. Although eating the same plants, these three species do not compete but complement each other in their use of the grass. While most of the herd feed heads down, several wildebeests stand sentinel on an old termite mound, each animal facing in a different direction for maximum coverage of the view.

regenerate at the expense of the grass, greatly reducing grazing not only for the hippopotamuses but for all the other grass-eaters as well. Since hippos are powerful animals with virtually no natural predators to keep them in check, the only means of preventing further pasture deterioration is extensive controlled cropping by shooting.

Cattle Under normal conditions ten or more species of large ungulates can live at high density over a wide area because they make use of the full spectrum of grasses, herbs, shrubs and trees. Their numbers are in delicate balance with the amount of plant growth, and even when food and water are short, over-grazing with consequent lasting damage to the pasture rarely occurs. In contrast, replacing the indigenous ungulates with equal, or even considerably lesser numbers of exotic species, as man has done over vast areas of Africa, may cause rapid and irreversible changes. The three types of domestic stock, cattle, sheep and goats, are changing the face of Africa as surely as they have already changed much of the Mediterranean region from lush forests and pastures into barren, stony hillsides.

Cattle select the sweeter grasses and feed by gathering the grass into their mouths and cropping

it off almost to the ground. They take in the stems which their digestive systems cannot utilize, but which could be used efficiently by the better adapted zebra and topi. Sweet grasses draw their moisture from roots a few centimetres below the surface, and when the grass cover is removed by cattle during the dry season, this moisture evaporates rapidly; under very heavy grazing and trampling, the nutritious grasses are killed. Sheep are then brought in to eat the coarser grasses and after them goats, to tackle what the cows and sheep will not eat. Goats can eat almost anything; they pull up plants to get at the roots and climb trees to strip the foliage. They effectively prevent all regeneration of grasses and palatable herbs. At the same time tough, poisonous or spiny plants get the upper hand through lack of competition from other plants. The herds on their way to water follow the same routes each day, and their trampling hooves wear gullies in the soil down which water rushes in the rains. Some gorges in East Africa, now over 10 metres deep, started as cattle trails.

Pastoralism is a way of life forced upon many peoples by the climate of the areas in which they live. Rainfall is often less than 75 centimetres a year, and this is too dry for most crops. At best it produces marginal grazing for cattle, yet here the pastoral peoples of East Africa, the Masai, Samburu, Suk, Watutsi and so on, have traditionally close-herded cattle, sheep and goats for subsistence. These peoples do not eat meat, but their live animals provide them with a staple diet of curdled milk and blood, and with many other necessities such as dung for hut building and for fuel. The cattle are not usually slaughtered until they are old.

The cattle-dependent races are splendid people, physical giants compared with the agricultural races who in the past were stunted by protein-deficiency diseases, to which their young children were especially vulnerable. But the problem for the pastoralists is that their livelihood depends on the quality of the grazing. Where grazing has deteriorated badly, only very small amounts of blood and milk can be taken from each animal. The poorer the grazing, the more head of cattle are needed to supply enough blood and milk for each person; and

the more cattle there are the worse is the over-grazing, so that each animal can supply less and less. Where peoples are dependent on cattle and have no other means of subsistence, they face starvation as the grasslands are turned into deserts.

The effects of over-grazing are more obvious and look worse in arid than in wetter districts. Today, in many areas where twenty or thirty years ago lush grassland supported game from elephants down, the end point of over-grazing has been reached: stony red soil barely covered by sparse thornbushes and spiky succulents, cattle skeletal even in years of good rain, and the game virtually gone. Ants and termites are sometimes blamed for damage to pasture because their nests can be more easily seen and appear more abundant in over-grazed areas; yet grasses and trees survived their onslaught for

A Tugen herdswoman carries a new-born kid as her goats graze towards the lake. Around the shore flat areas are still grass-covered, but the hills behind have been grazed and eroded to bare stony ground.

Over-grazing by goats has reduced this area of bush to bare stones in which only spiky and unpalatable plants persist—aloes, a patch of grey-green *Caralluma russelliana*, bayonet aloes and acacias. It is now the rainy season, yet no grass or seedlings sprout. How the hinged tortoise survives is a mystery.

Cattle assist erosion in various ways—here rather unusually by eating mineral-rich earth from the sides of a shallow ravine. There is little enough grass left in the area for them to eat, as the bare earth above the cliffs shows.

millions of years until man and his cattle destroyed the balance. In the end, even the ants and the termites may die out from these areas.

Cattle have been in Africa for four or five thousand years but they have failed to adapt fully. They suffer from heat stress and need to drink at least once a day; and endemic diseases cause a heavy death toll. The most serious of these diseases is *trypanosomiasis*, or sleeping sickness, known as nagana in cattle. It is transmitted by the tsetse fly. The flies feed by piercing the skin and sucking blood until they are bloated. While doing so, single-celled protozoa—trypanosomes—are transmitted to the host in the saliva of the fly. These undergo part of their development in the blood of the host, part in the gut of the fly.

Tsetses are endemic to Africa south of the Sahara and there are over twenty species. Some live in thick forest, some in bush, some near lakes. Wild mammals, birds or reptiles, which evolved together with the tsetse over millions of years, do not succumb to the diseases which they carry, but introduced cattle do, and so are impossible to keep in a third of the continent, wherever tsetse flies occur.

In some parts of Africa from Uganda southwards, man has attempted to eliminate the tsetse from potential cattle areas by shooting out the game which acts as a reservoir for the trypanosomes, by felling trees and bush that harbour the insects, and by trapping the flies and spraying the countryside with insecticides from the air. Such operations cost millions of pounds and are only marginally successful. Despite the shooting of many millions of big animals, small and elusive game such as bush pigs and duikers escape the slaughter and continue to act as reservoir hosts for the protozoa. Dense thickets of thornscrub invade the cleared bush, so that the land, even if it is free of the tsetse, becomes useless for grazing cattle. The policy, so terribly costly in natural resources, was a failure and has been suspended. Nowadays, the tsetse is looked upon by conservationists as the saviour of Africa, conserving the soil from over-grazing and erosion. At the same time the environment and the wild ungulates in it are acknowledged as a unique natural resource, to be utilized wisely, as well as enjoyed for the pure pleasure of seeing beautiful animals in the superb unspoilt African countryside.

The great beasts of prey The big herbivores are in constant fine balance with the plants of the environment. In just as fine a balance with the herbivores are the great carnivores. The chief predator of plains game is the lion, and its prey is generally the dominant herbivore on its home range. In some places this may be buffalo, in others wildebeest or zebra; it does not show a particular food preference, but usually takes what it can get most easily. A lion can hunt by day but more often does so by night. It employs various techniques, sometimes hunting singly, sometimes in groups. In thick cover a lion will stalk its prey by stealth, creeping belly to the ground, freezing when the antelope's head comes up alert, moving only when the antelope lowers its head again to feed. In the open it will move purposefully after its prey, quickening its pace as the animal turns to flee; then with a powerful spurt of speed it gains on the antelope and bowls it over. Sometimes lions lie up and wait in dense cover, often near a waterhole among bushes which the game must pass when they come to drink. At times a lion will show itself or roar or give its scent on purpose to stampede the prey towards other lions waiting in ambush.

When seriously alarmed, plains game flee at full speed, high jumping in every direction. Impalas, particularly, do most spectacular leaps. This serves to confuse any predator whose hunting technique requires concentration on a single individual, and also enables the fleeing antelopes to keep watching the enemy as they bound through bush cover. The black and white markings on the sides and rumps of many antelopes, and the stripes of zebras, accentuate the dazzle effect as the animals streak about.

The largest purely diurnal predator is the cheetah. A sprinter, and the fastest animal on land, it fetches down its prey by sheer speed. It may take any one of twenty-five different prey species, but on the plains Thomson's gazelles are its main food, while in bushy areas it takes chiefly impalas. Like a lion, a cheetah selects a particular animal before trying to catch it. It may stalk its prey for up to half an hour, with head and body low, taking advantage of every termite hill or bush to cover its approach. When it is close enough for the final rush, the cheetah dashes out, dodging and zigzagging after its fleeing prey until it can bowl it over and grab it by the throat.

The third of the great African cats, the leopard, is the most secretive of the three. It is usually found in forest or wooded places, and takes its prey by leaping on it from a branch or from its ambush in

Part of a vast herd of Samburu cattle returning home to their village in the evening. The rainy season is well advanced yet the cattle remain pitifully thin. In a few years' time this trail may become a ravine down which flash floods briefly roar during the rains. It is not the pastoral system itself which damages the habitat, but keeping more cattle than the grass can sustain.

A leopard draped along a branch at sunset with its tail hanging down like a bell-pull. It will stay feeding through the night on its kill, a waterbuck calf, which it has dragged up the tree out of the reach of vultures, jackals and hyenas.

Three impala bucks of different ages pause to nibble grass as the bachelor herd crosses the dusty bed of a former lake. In contrast to the poor-looking cattle they are beautifully sleek; by feeding on a variety of leaves and twigs as well as grasses they are able to utilize an arid habitat to the full. In this tsetse area they are probably infected with trypanosomes, but unlike cattle they will not succumb to the disease they are carrying.

thick cover. In the forest it feeds mainly on duikers and other small antelopes; up mountains it lives on hyraxes; among rocky hills and cliffs it takes baboons and klipspringers. It also preys heavily on the newborn young of plains game.

It has been found that leopards more often kill male animals than females, except in the breeding season when the very young are taken indiscriminately. The same is true of lions and cheetahs: their prey is more often males in their prime than females or yearlings. Among plains game, adult males make up a fifth of the population, yet of kills made by the big cats, three-fifths are adult males. This is because males are easiest to catch, being more likely to be separated from the herd. In many antelopes a lone individual is almost invariably a subordinate male who feeds by himself while the females and young

stay together. In others, such as kob and Thomson's gazelle, the solitary males are the successful territory owners. Bachelor Grevy's zebras are solitary and territorial, while the family herds graze in separate areas. Animals in herds rely on many pairs of eyes, ears and nostrils to detect danger and so may escape a predator whereas a lone male is more readily taken unawares.

The effect of predation is therefore to reduce the number of males. In a harem society, however, only a few males are needed to maintain the population. The excess males promote healthy competition and ensure that only the most virile sire the next generation. They also provide food for the carnivores, who thus keep the herbivore populations stable. Plains game are capable of enormous increases in their populations if unchecked.

Left: The cheetah is the fastest animal on land and preys on gazelles and antelopes. It does not hunt for itself alone, but leaves much of its kill for the scavengers. Replete, it suns itself, idly scratching.

A bachelor Grevy's zebra patrols his territory in the early morning; mist and cloud are low on mountains of the Matthew's range. Solitary adult male herbivores are most vulnerable to predation by the big cats, and make up over half of their kills. When this picture was taken, the stallion on the adjoining territory had just been killed by lions (above).

A young lioness, her belly distended with meat, half-heartedly gnaws at the remains of the Grevy zebra stallion. Her pride made the kill before dawn; now it is evening and they have been feeding all day. The other lions lie gorged and panting nearby.

Although they give birth to only one young a year, they can multiply at a rate which can double the population every four years. If they are not preyed upon, they soon outgraze themselves, so predation is in fact an essential process in preventing over-population. The hundred herbivores a pride of lions kills in a year represents only the annual breeding increment and under normal predation, in normal climatic conditions, the herbivore populations remain remarkably constant.

But what controls the numbers of the predators? To some extent, especially among lions, territorial behaviour may be important in limiting numbers. Unlike the ritualized combats between rival antelope males, which are usually resolved without serious injury to either contestant, war between territorial male lions is often a fight to the death.

Also, predators breed at a rate that keeps pace with the food supply. If the plains game decline, mortality among lion cubs may be high, as many as half a litter being lost in the first year. But in places where game is abundant, mortality of cubs is very low. If the buffaloes, zebras and antelopes increase, then lions rear more young and therefore need to kill more prey. Thus an equilibrium is maintained.

However, the delicate balance between herbivores and carnivores is never simple, especially in an environment so rich in species of both predators and their prey. On the plains the situation is complicated by the presence of two other big predators, hunting dogs and spotted hyenas; all five major predators may prey on the same big grazers.

The hunting technique of hunting dogs is quite different from that of any of the cats. Hunting dogs are long distance runners, with impressive staying powers. They are not built for great speed, yet have sufficient to run down any antelope. They hunt in packs, and after their leader has selected the prey, they follow it until it begins to tire. The lead dog then increases speed, catches up with the animal and further exhausts it by repeatedly snapping chunks from its belly or from any part it can grab. Finally the animal either falls exhausted, or several

Hunting dogs, the most efficient killers of the plains, are gregarious and hunt in packs in an organized way. They play an important part in the natural balance, controlling the numbers of herbivores like any other predator; and like the big cats they can seem lethargic in the heat of the day.

The web of life in grassland. Grasshoppers, grass-eating rodents, and plains game large and small are the bases of the three great food pyramids of the plains. In complex natural communities very few food chains are simple; very few animals avoid falling prey to other animals. Most grass-eaters are preyed on by either insect-eaters or by carnivores; carnivores may eat the insect-eaters and the carnivores may in turn be eaten by larger carnivores. Only the great super-predators and scavengers at the summit of the pyramid are free from the threat of predation. This diagram does not attempt to show all the animals involved in the intricate grassland pyramid, but illustrates representatives of the various groups.

dogs bring it down and rip it to pieces. Small antelopes are killed very quickly, but larger beasts take longer to die, though their death is not necessarily any more lingering than if it were inflicted by a lion or a cheetah. In fact, hunting dogs are the most efficient killers of all the large predators. Once the prey is selected, it does not escape. The majority of antelopes killed by hunting dogs are, again, territorial males, and because the dogs are nomadic, the herds are never harassed unduly, or over-cropped by them.

Hunting dogs kill by day; spotted hyenas by night. Hyenas are not merely scavengers, but are

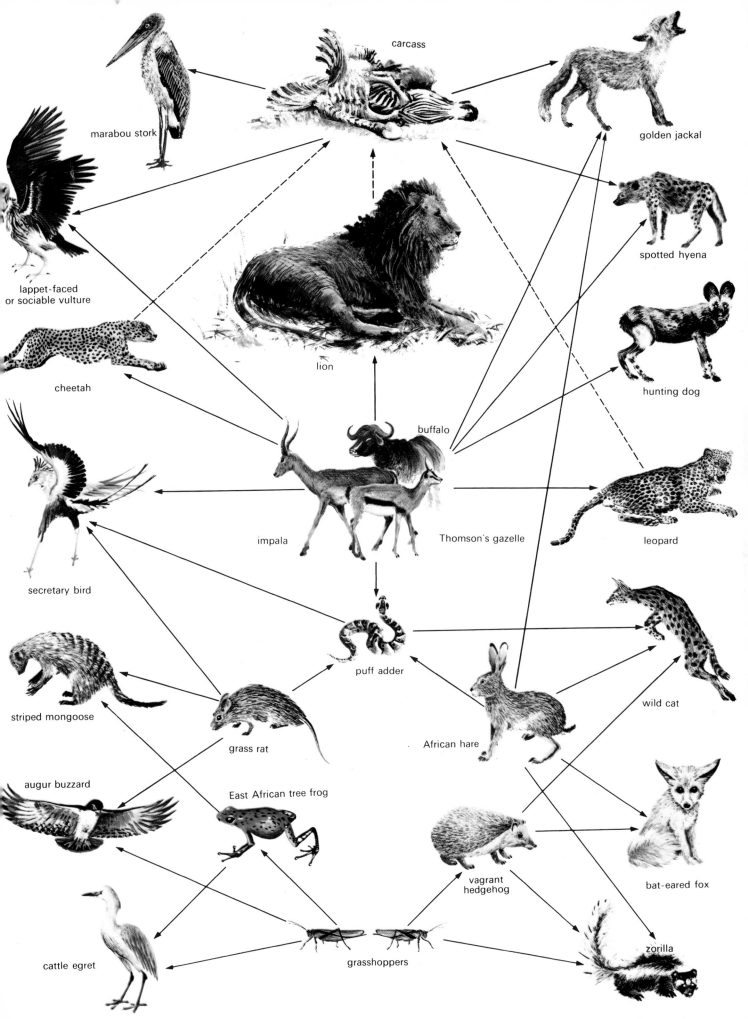

marabou stork

carcass

golden jackal

lappet-faced
or sociable vulture

spotted hyena

cheetah

lion

hunting dog

secretary bird

buffalo

impala

Thomson's gazelle

leopard

puff adder

striped mongoose

African hare

wild cat

grass rat

augur buzzard

East African tree frog

vagrant
hedgehog

bat-eared fox

cattle egret

grasshoppers

zorilla

The first vulture to arrive at the old lion kill at daybreak was this hooded vulture, lighter and more mobile than the griffons. When it began to feed, the bare skin of its head was quite pale. A little later griffons started to drop from the nearby trees and the hooded vulture blushed in defensive display.

also major predators in their own right. After dark a pack will set off in a seemingly leisurely way until prey is sighted. They begin to harass the herd until they can single out a sick or very young animal. The whole pack then concentrates on this and quickly brings it down, tearing it to pieces. Very little is left of the carcass, for hyenas have the most powerful jaws in proportion to their size of any living animal.

All the major carnivores, apart from hyenas, may leave pickings from their kill. There is meat or offal or bones enough to feed a variety of scavenging animals as well as the predator that made the kill. Cheetahs eat only the meat of their prey, leaving all the rest—skin, bones, digestive tract and its contents —for the scavengers. Leopards take their kills up into trees but when they have finished eating, arboreal scavengers such as crows and ravens, even bataleur eagles, pull the remnants about and bring them to the ground. There hyenas, vultures, even ants and other insects, finish off the hide and bones. Where a lot of large animals have been killed, there may not be enough hyenas to dispose of all the bones. On the route of the wildebeest migration (page 95) many bones and horns remain scattered above ground. Some rodents such as rats and porcupines may gnaw at them, but in the end weather and bacteria break down the bones and specially adapted fungi grow on the horns.

Hyenas are the main scavengers by night, but during the day their place at the carcass is taken by vultures, which depend on lions to provide most of their food. There are ten species in Africa, six of which are common in East Africa and may even feed together around the same carcass. Although competing for food at times, the species are generally separated by their preferences for different types of country. Rüppell's griffon vulture is a bird of dry bush and desert; the white-backed vulture prefers grassland dotted with acacia trees. The former nests on cliffs; the latter in tall riverine trees. All vultures cover such vast distances in their search for food, it is scarcely surprising that several species sometimes converge on the same piece of carrion.

Vultures find their food mainly by soaring high in the thermals and watching not only the ground for lions or their prey, but also the behaviour of other vultures. Cruising effortlessly hundreds of metres above the ground they can spot another vulture dropping down a long way away. Other vultures even further off in turn see them descend, and so all soaring in a wide radius may converge when a kill is discovered. Animals on the ground also watch vultures descend, especially jackals and hyenas which sometimes locate food this way. Sometimes vultures perch in dead trees, keeping watch over lions, hoping they will rouse themselves and make a kill. Or the lions may already have a kill nearby, which they have dragged into the shade and are guarding. Vultures often have a fruitless vigil, driven off by lions, jackals, hyenas, even marabous, whose long reach and huge sharp bills give them an advantage. When vultures alone are at a carcass, the hungriest has precedence. Any famished new arrival puts on a great display at touchdown, bounding towards the carcass with neck arched and wings spread. The less hungry birds that have already bolted some meat give way before this apparition.

However unlovely vultures may appear, they are beautifully adapted to fill their role of scavengers. Their long sight and powerful wings often bring them to carrion that may not be discovered by terrestrial scavengers. Their function is therefore complementary to that of jackals and hyenas.

The plains, by virtue of their openness, offer a unique opportunity to see at a glance the basic relationships of what is a complex ecological system. The herbivores can be seen consuming vast quantities of grass while the carnivores that keep their numbers in check lurk in sheltered places. Overhead, vultures wheel, while perhaps a solitary jackal trots purposefully along a well-worn trail. The overriding impression is one of harmony, with each animal perfectly adapted to fit a particular way of life.

Just after dawn, griffon vultures descended in a squabbling mass onto the old kill, driving off the small hooded vulture. Vultures are an essential part of the ecology of the plains, rapidly and efficiently disposing of carrion during the day.

The black-backed jackal feeds on any small game, and as a scavenger is also attracted to lion kills. This jackal is lying near the remnants of an old kill just before dawn. It is fully fed and is making way for the hungrier vultures; it will soon trot along home with its mate to their den in an old ant-bear burrow.

The Dry Season

East Africa is essentially a hot place. Even in the highlands, the dry season daytime temperature may be 27°C. On the plains, midday temperatures may exceed 38°C. During the dry season the sky is often cloudless throughout the day, or with only wisps of high, dappled clouds in the afternoon. When humidity is low, more solar heat gets through to the ground in the daytime, and more is lost from it during the night. So in the cloudless dry season, with the sun vertically overhead at midday, the heat may be intense, and the nights that follow correspondingly colder. In the heat of midday the scenery quivers and vibrates, and silvery mirages appear like lakes above bone-dry ground. Moving game in the distance seem to be wading through the shimmering water. The sizes of animals are distorted by heat haze and distance, so that gazelles may look like buffaloes, or a jackal like a lion. The sense of intense heat is accentuated by the faded colours of the landscape; the dead greys of leafless thorn bushes, the pale gold of dry grass, with blues and purples in the distance.

At this season there is often a dearth of small animals. Butterflies are few; grasshoppers, small spiders, aphids, even ants, scarce; caterpillars apparently non-existent. Toads and invertebrates that are very vulnerable to desiccation, hide during the day in the relative coolness and humidity of the earth beneath logs and stones. Other animals go to ground for the duration of the dry season and pass it in summer sleep or *aestivation*. At the start of the dry season, these animals burrow into the ground before it becomes too hard. During aestivation, their metabolism is greatly lowered, as it is in animals that hibernate through very cold weather. Tortoises and terrapins aestivate, as do some frogs and very many invertebrates, and the fat mouse probably also does. Giant snails cement their shells shut with mucus which dries into a hard cap to reduce evaporation.

Among East African animals that aestivate, one of the most unusual is the lungfish. As soon as its swampy habitat begins to dry out, it buries itself in the mud, curls up and secretes a mucus which hardens as the mud dries, to form a cocoon in which the fish's body stays moist. Lungfish, as their name implies, possess primitive air-breathing organs which correspond to the lungs of higher

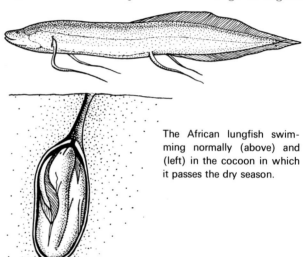

The African lungfish swimming normally (above) and (left) in the cocoon in which it passes the dry season.

Previous page: The skull of a greater flamingo lies on the cracking mud. During the dry season lake waters shrink rapidly; puddles, waterholes and even rivers dry up, leaving much of East Africa a waterless wilderness.

Below: A Thomson's gazelle ram on the parched plains seems to be making its way towards a shimmering lake. Doubtless it is not deceived by the mirage, since an antelope depends as much on its nose and its memory as on its sight to guide it to water.

vertebrates. No other living fishes possess true lungs, although some breathe air by other means, modifications of swimbladder, gills or gut. Lung-fishes still have gills, though much reduced, with which they can breathe in water. While buried in the mud the lungfish is able to go on breathing through perforations in the top of the cocoon, and so exists until rain releases it.

The African catfish also buries itself in the mud during the dry season, as it is one of the fish with accessory air-breathing organs. In swamps and other poorly-oxygenated water it surfaces frequently to gulp a mouthful of air and instantly shoots down to the bottom again. It is extremely wary, and under non-drought conditions would not be caught napping at the surface by fish eagle or heron. When the pools begin to dry up it is quite comfortable breathing air and guzzling up the less fortunate smaller fishes stranded gasping and flipping in the shallows. Extremely voracious, it is capable of such an intake of food that its belly becomes spherical, but in shallow water it is just as likely to be taken itself by all sorts of opportunist fish-eaters. Indeed, in some districts fish eagles time their nesting to coincide with an abundance of catfish in dry season low water.

Water birds

Other large water birds besides the fish eagle nest in the dry season and hatch their young when

The lesser flamingo nests at the beginning of the dry season when receding water exposes mud flats in ideal condition for nest-building.

Kittlitz's sand plover may nest on sandbanks or lake shores during the dry season to avoid the danger of floods. When the bird leaves its nest it does a little dance around it, scuffing sand over the eggs to hide them and to shield them from the hot sun.

lakes and rivers are drying up and aquatic prey of all kinds is easiest to catch. Herons, egrets, storks, spoonbills and ibises nest in big mixed colonies in trees, usually on an island or lake shore, sometimes far from water. Smaller birds that nest on the ground near water and so are vulnerable to flooding, often breed when the waters are receding. Gulls and terns, plovers, skimmers, pratincoles and waders nest on sandbars and shores. In the dry season, too, the receding waters of the soda lakes expose acres of wet mud which lesser flamingos need for breeding on.

Lesser flamingos breed chiefly on Lake Natron, in enormous colonies 12 kilometres out on an alkaline mudflat, hidden from the shore by mirage. Another breeding site in East Africa is a crater lake on a small volcanic island on Lake Rudolf. On Lakes Nakuru and Hannington the mudflats are not isolated from the shore, and though the flamingos build nests they do not breed successfully. The nests are pedestals of mud which the birds scoop up with their bills. On the baking mudflats the midday temperatures are high, but the long legs of adult flamingos carry them above the worst of it and on the top of their mound nests the temperature is only at blood heat. The disadvantages of high temperatures and glare are outweighed by the advantages of isolation from most predators.

In very hot areas such as Lake Rudolf crocodiles leave the water before sunrise and come out to bask. Almost all crocodiles come out onto the shore, except the big territorial males who stay in the water, lying half-submerged or patrolling in the shallows. When the sun becomes too fierce, the basking crocodiles retire to the shade or go back into the lake. By following a daily pattern of alternating periods in the sun, in the shade, and in the water, they are able to maintain a body temperature varying only a few degrees above or below 25°C. They also regulate their temperatures by basking with the mouth open, to radiate heat and evaporate moisture.

The heat of the day

During the hot dry season the majority of big game animals spend the heat of the day resting in the shade of bushes and trees to avoid overheating and dehydration. Elephants feed in the early morning and again in the evening as they make towards water, but from about mid-morning until mid-

afternoon they remain quietly in the shade, standing beneath a thorntree or occasionally lying down. Black rhinoceroses, too, may stand beneath thorntrees in their characteristic dejected-looking, head down attitude, relying on their attendant tickbirds to alert them to approaching danger. Warthog males spend the greater part of the day lying up in the shade of bushes, but females and young must spend more of their lives feeding, so only lie up during the very hottest part of the day. The exuberant savannah chimpanzees fall quiet and doze through the heat of midday.

Most of the large herbivores rest in shade during the heat, usually standing, but sometimes lying down with head up while they chew the cud. Ruminants rarely lie on their sides as this position interferes with the smooth working of their complicated digestive systems. A few antelope, such as Bohor reedbuck, make form-like shelters in tall grass where they lie up like hares. Impala rest together among bushes, backs to the breeze, most of the herd lying down, with some older females standing to keep watch.

While herbivores are resting in the shade, they may shut their eyes but they do not often sleep. Their rhythmically cudding jaws, and the twitching of ears and tails, show they are only lightly asleep. Only occasionally does an animal go into a brief, deep sleep. A lightly sleeping antelope remains with its head up but, in contrast, one going off into a deep sleep stretches its neck along the ground and allows its head to flop over. Antelope do not sleep deeply for longer than four minutes at a time, but zebra may stretch out and sleep soundly for ten minutes. During the night zebras lie down together to sleep, with one or two standing sentinel. During the day, however, they do not sleep communally; an individual will lie down and sleep by himself, with zebras or other plains game standing watchful close by.

During the heat of the day the big cats appear most lethargic. Leopards loll along high branches of tall trees, lions collapse in the shade of bushes, especially near half-eaten prey, which they may continue to guard from scavengers. A lion that has gorged itself often looks uncomfortably hot, not only at midday but in the cool of the evening as well. This is because he must pant heavily to lose water so as not to become overheated. Man and

Right: A pair of Masai giraffe resting and cudding at midday in the shade of feverthorn trees.

horses, including zebras, have well-developed sweat glands, but most other mammals, including lions, have few. Hunting dogs have sweat glands in the feet, so cool themselves by evaporation from the pads as well as by panting and dripping saliva from the tongue. The ears of some animals are secondary devices for radiating heat from the body: elephants constantly fan their great ears backwards and forwards, and antelopes flick their ears rapidly.

Birds fare rather better in a hot climate than do most mammals, since their body temperature is higher. Whether they inhabit the tropics or the arctic their temperature is 41°C. when inactive, and up to 43°C. during intense activity. Nevertheless they occasionally need to lose heat, and, having no sweat glands they, like many mammals, must pant rapidly to cool themselves. Some species pant at the rate of up to 300 respirations a minute, and like the lion, look as if they are in distress from the heat.

An example of the way birds can withstand high temperatures is the freckled nightjar (page 61) which lays its eggs on completely exposed bare rock. At midday the dark-coloured rock absorbs all the sun's heat and feels baking hot to the human touch; yet the incubating bird manages to maintain its eggs at the right temperature for development.

The animal that is best adapted to life in hot places is the camel, a true desert animal kept by tribes in the northern districts of Kenya but not brought much further south than the semi-arid country around Lake Baringo. The camel is able to lose heat by sweating, but need not do so except in extreme heat because its body temperature can rise several degrees without harmful effects.

Grant's gazelles and Beisa oryx also allow their body temperatures to rise by up to 6°C. during the day, from about 39°C. to 45°C., before they start sweating. Eland may allow a three-degree temperature rise. Such antelopes have a special system for cooling the brain more than the rest of the body, to avoid brain damage which would otherwise result from high body temperatures. Grant's gazelles are therefore able to remain in the open during the hottest weather, while most other animals seek the shade. They have very shiny coats of short hair which reflect the heat, as do goats' glossy hard coats and the iridescent plumages of starlings. Their very pale sandy coloration may also contribute to heat regulation by reflecting much of the sun's radiation. On the other hand,

the pale coloration may be mainly protective; Grant's gazelles exactly match in tone and colour the yellow plains grass in the dry season.

It was once argued that pale coloration was an adaptation to a hot environment because it reflected heat, while dark colours that absorb heat were thought to be an adaptation to a temperate climate. But domestic sheep in semi-desert places may be black, white or pied, or strikingly coloured with foreparts white and hindquarters black. Zebras, too, are black and white, reflecting and absorbing in almost equal amounts. (The zebra's stripes are, however, generally regarded as a protective device which renders the animal almost invisible at dawn or dusk or in moonlight; and as a dazzle-effect for confusing predators at close range.) Even dark-coloured wild animals such as wildebeests, buffaloes, rhinoceroses and elephants do not always seek shade at midday. Buffaloes spend most of the day resting and cudding in some peaceful place. Most often this will be in the densest cover they can find, but not because they need the shade. Their chief requirement is tranquillity, so that in places where buffaloes are not persecuted, their resting place, except in the very hottest weather, may well be in the open.

Some black rhinoceroses show a marked preference for dusty depressions and sand bowls as resting places, and they, too, will remain out in the full sun. A rhinoceros has to get up about every

A rapidly panting small bird can look as if it is suffering extreme distress from the heat, but a large bird like the yellow-billed stork simply seems to be enjoying a joke. Lesser flamingos achieve some heat-loss all the time, paddling in the cooler lake.

The camel is a true desert animal and looks a little incongruous among the comparative verdure around Lake Baringo. A large youngster follows its mother closely, holding onto her tail.

Right: A small herd of wildebeests grazing the parched yellow plains. The grass looks dead, but is still nutritious and will sprout afresh as soon as rain falls. In spite of their dark coloration, wildebeests do not suffer from heat-stress and do not need to seek shade except in the very hottest weather.

The pale sandy coloration of the Grant's gazelle blends with the dry grass. It is one of the few mammals that habitually stays out in the open under the powerful midday sun.

Black rhinoceroses sometimes rest out in the full midday sun—here in the baking heat of a vast dustbowl whose fine, whitish alkaline soil was once part of a lake bed. The pair appear to be asleep, but are only dozing. A call of alarm from their attendant oxpecker would alert them instantly.

hour and a quarter while he is resting, to relieve the cramped sleeping posture for ten minutes or so, but it will remain in the same dust bowl, alternately lying down and standing up, for ten hours at a stretch. A resting rhinoceros' great hulk blends surprisingly well with the dull grey thornbush scrub, especially when both animal and vegetation are coated with a layer of the same dust. The small cloud of dust which the rhino raises as it exhales is often the only thing that gives away its position when it is in thick bush.

Living without rain

Leafless thornbushes offer little shade, so poorly-adapted animals must endure the heat as best they may until the evening brings coolness. Then they become active again, making their way towards water, which means not only a drink but a heat-dissipating wallow in the mud. Rhinoceroses, elephants, buffaloes, warthogs and spotted hyenas wallow as an efficient method of cooling the body and disposing of heat accumulated during the day. A wallowing animal lies in the mud to plaster each side of its body, and may even roll right over to plaster its back as well. Ostriches, which often

behave more like ungulates than birds, will sometimes wallow in water, submerging all but the neck and head. Zebras and antelopes do not wallow; perhaps this is a luxury enjoyed only by the largest or best protected animals. The majority of herbivores come to water only to quench their thirst, and do this as quickly as they can. Impala come in a close herd; each animal drinks from the very edge for less than half a minute, some remaining watchful while others drink. Then they and other plains game leave as quickly as they came, for at the waterhole they are especially vulnerable to all kinds of predators, from crocodiles and pythons in the water, to leopards and lions in the bushes.

In contrast to the hastily sucking antelopes and zebras, lions take their time at a waterhole. Their method of drinking is slow and laborious; it may take twenty minutes for them to get their fill by lapping. Lions prefer clean water from a flowing

During the dry season, game comes to water at any time of the day, but mostly in early morning or evening. Here wildebeest drink at sunrise as three impala bucks depart.

river, but they will drink from a stinking stagnant waterhole if necessary in the dry season.

The majority of animals come to a waterhole during the afternoon and evening, especially when food is plentiful and the weather not too hot. But when food and water are scarce, feeding grounds for elephants may be more than 30 kilometres from water, so that the animals cannot return to drink until late at night or even until the middle of the next day. Each adult elephant consumes 140 to 170 litres of water, sucked up in the trunk 9 litres at a time, and squirted into the stomach with a noise like a cistern emptying.

It is obviously a great advantage to plains game to be able to make up water-loss quickly. The donkey is able to make up a water-loss of twenty per cent of its body weight in less than two minutes. Other dry country ungulates are probably able to do likewise. The camel needs to visit water very infrequently, and can exist for a fortnight on only dry food. It can tolerate a much greater depletion in body weight than most other mammals, more even than the donkey, and may lose thirty per cent of its body weight without ill effect, compared with twelve per cent for most other animals. At the same time its appetite does not diminish

Warthogs at a wallow in the late afternoon. The smaller one is enjoying a muddy drink, while the larger one cools itself by rolling luxuriously, coating itself all over with wet mud. As well as cooling, the wet mud may act as a soothing plaster for minor sores and irritations, and a deterrent to biting flies and other parasites. When dry it is an effective camouflage among termite hills and dusty bush.

with desiccation until this becomes severe. It is also able to make up the water-loss afterwards by an intake of water such as would kill most other animals.

The adaptation of the camel and donkey to a dry environment does not involve complete independence of drinking water, but rather an ability to economize in its use. Oryxes and Grant's gazelles are tolerant of very dry conditions, and are able to stay for long periods in waterless country during the dry season—an advantage that enables them to feed over large areas not available to the water-dependent species. Drought-tolerant ungulates like these obtain most of their water intake from their food (the dryest grasses found under natural conditions in East Africa contain three to five per cent water by weight) and lose very little water in their sparse and concentrated urine and droppings. Browsers get all the moisture they need from the leaves of bushes, which are hygroscopic and which, even during a drought, can consist of sixty per cent water. Antelopes feed mainly at night, when relative humidity is higher and leaves are wet, and they may obtain additional water by licking dew. In arid regions, where plant cover has not been lost, the volume of dew may be greater than rainfall. In one very dry area, 40 kilometres away from the nearest free water, impala drank dew in the early morning and went on drinking it in the shadow of trees until it had all evaporated. But in dry, over-grazed areas there is no appreciable dew because evaporation from the bare soil is too great.

A few animals are able to live in very arid bush without ever drinking standing water at all.

A small gerbil manoeuvres a stone backwards out of the entrance to the burrow it is digging. Gerbils dig rapidly into sandy soil, throwing the earth out with their hind feet. They spend the day underground to keep cool and conserve moisture.

An extraordinary rodent, and one of Africa's oddest mammals, is the naked mole rat. As its name implies, it has no fur, just a few tactile bristles; and no ear-flaps.

Gerenuk, for example, have never yet been seen to drink in the wild. Gerbils also inhabit arid bush and semi-desert, and exist entirely without drinking. There are many gerbil species, ranging from very small mouse-size to large rat. All are nocturnal, sand-coloured, with big eyes. They are related to rats, but intermediate in appearance between true rats and jerboas, very specialized true desert-dwellers not occuring in East Africa.

Gerbils are much more mobile than rats, and probably lead a more nomadic existence. They subsist mainly on seeds, supplemented by bulbs,

Like most other birds, the brimstone canary needs to visit water regularly to drink. It drinks by alternately sipping and raising its bill, tipping the water down its throat.

Scorpions are common in dry country. During the day they remain in the coolness and humidity of a burrow or beneath a stone; when they emerge at night they use their extended pincers as feelers to find their way and locate small animal prey.

tubers and insects, all of which contain a fairly high proportion of water. Even the driest seeds, like the driest grasses, contain some water, and more is formed by the oxidation of food during metabolism. However, physiological economy of water is very necessary, so the animals must stay underground throughout the day. All species of gerbils are therefore inveterate excavators, often digging mazes of burrows about 15 centimetres below the surface.

Gerbils are the prey of nocturnal predators, especially those whose approach is stealthy, such as owls and small cats. Snakes may seek them in their burrows, and monitors try to dig them out, but a gerbil is often able to escape underground by digging further into the sand and closing the burrow behind it, or by bursting out of a stopped-up bolt-hole and running swiftly to another set of burrows.

Another small mammal of some very dry regions is the extraordinary naked mole rat. This lives entirely underground, only occasionally coming to the surface, perhaps to collect seeds or other vegetable foods, though normally it feeds on tubers, bulbs and roots underground. It is a baby rat-sized rodent; indeed it looks like a baby animal for it has no fur, only a few sparse bristles, a fringe of hairs around the edges of its feet to help it shovel fine soil, and whiskers on its lips.

Most burrowing mammals, for example, the golden moles and mole-shrews of the mountains, and the well-furred Kenya mole rat, live and work alone, but the naked mole rat is unique in that it lives in colonies of up to a hundred individuals which all co-operate in the digging of underground tunnels. Teams of naked mole rats form living chains, with three distinct working sections. The excavator does all the digging, biting at the earth with its big incisors, gathering it with the forepaws and kicking it to the animal behind. This animal collects the soil from the excavator, and then works its way back along the passage, beneath the rest of the chain gang. It leaves the load of earth with the final animal at the tunnel entrance and rejoins the excavator by crawling over the backs of the other members of the chain gang, who are moving backwards in their turn with loads of earth towards the entrance. The soil-disposer at the burrow mouth vigorously kicks the earth out, straight up into the air. The soil is too fine and dry to form the usual type of mole-hill, so the spurts of dust fall to form typical naked mole rat volcanoes, that can be seen apparently puffing smoke when the excavators are working busily.

Several invertebrates are found in arid areas and exist without drinking. Under nearly every stone a scorpion may lurk. Dry country scorpions are usually rather small and finely-built, with brown bodies and yellow, almost translucent pincers; forest scorpions are often great greenish-black creatures with massive pincers. Desert scorpions avoid desiccation by remaining below ground during the day, and can exist for very long periods without drinking or eating. But when they feed they can gorge themselves to double their

The leg-like pedipalps of a solifugid are covered with long, fine, tactile hairs. Like the scorpion, this dry country animal feels its way about at night. Its jaws are the most fearsome of any invertebrate; the two pairs, side by side, work backwards and forwards to pulp up the prey.

The ground hornbill is one of the few birds known to be able to go without drinking for long periods. A bird of open country, it walks about on the ground, usually in family parties, snapping up dung beetles, millipedes, grasshoppers, lizards and fallen fruit.

A mother warthog with four piglets and a companion at the salt-lick. Warthogs prefer to eat soil when it is dry; some animals, such as buffaloes, prefer it when it is wet.

fasting circumference in a very short time.

Solifugids are another predatory invertebrate common in arid areas. Solifugid means "fleeing from the sun", and most species are strictly nocturnal. They, like scorpions, are wholly carnivorous, feeding on spiders, insects, smaller solifugids and scorpions, as well as mice, birds and lizards. They are rather horrific, reddish-golden, spider-like creatures, covered with long, stiff, tactile bristles, but they are not poisonous. They lack the scorpion's pincers for holding and its sting for paralyzing, but have two powerful pairs of jaws set side by side which work backwards and forwards to pulp their prey. Solifugids never need to drink; they obtain sufficient moisture from their food.

Among birds very few, if any, can exist entirely without drinking. Ostriches can go without water for many days, but they drink frequently when it is possible. A ground hornbill is known to have gone six months without drinking. But most birds, like most mammals, are not fully adapted to a dry environment, and must visit water frequently to drink. In arid areas birds fly long distances to water. Flocks of pigeon-like sandgrouse flighting to and fro in early morning and evening are a familiar sight in semi-desert country. The birds may gather in huge flocks to drink, some having flown up to 64 kilometres to the waterhole. Here they gulp enough water to last them for the day. During the breeding season the cocks take water back to the chicks by means of a remarkable modification of the breast feathers which enables them to soak up and retain twice as much as normal feathers can carry when wetted. At the waterhole a cock sandgrouse walks belly-deep into the water, then crouches down and ruffles out his feathers, rocking to soak them. Even after flying twenty miles in hot weather he can deliver about half an ounce of water to his chicks, who nibble the feathers through their beaks to extract it.

Dust

Some birds bathe in water, not so much to cool themselves as to help maintain their feathers by the preening and oiling that goes on afterwards. Other birds never bathe in water. Instead they bathe in dust. Guineafowl, francolin, bustards, sandgrouse and mousebirds flap and ruffle their feathers in it, even roll in it, kicking it over themselves with their feet. Afterwards they explode into the air in puffs of dust, like little rockets. Ostriches also dust-bathe, and raise a small dust-storm when they shake their feathers afterwards. Lammergeiers crouch in dust, and their breast feathers become stained according to the colour of the earth. Those stained russet with iron oxide were once thought to belong to a separate race from lammergeiers with paler breast feathers. These are all birds of open places where dust is more abundant than water. Others may dust-bathe at salt-licks or under overhanging rocks, anywhere where dry earth is accessible. Here hoopoes hollow out dusting holes in which they rotate till they are half buried; larks shuffle around with drooping wings and ruffled plumage; rollers, bee-eaters, hornbills and nightjars wallow in the dust.

Dust-bathing is not confined to birds; some mammals dust themselves. Gerbils, particularly, like to roll in fine loose sand, and groom themselves afterwards. Members of the horse family all roll. A herd of zebras will visit a dusty spot and repeatedly roll, kicking vigorously. Elephants also make great use of dust. After drinking they spray themselves all over with water, or they may wallow. They then very often take a dust-bath as well, spraying their bodies with fine earth, so that, like the lammergeier, they take on different colours according to the soil of the region. In areas of fine white alkaline soil they may be pale grey; in the black cotton areas dark; and bright red on laterite. These coatings of dust may have a camouflaging effect during the dry season, but in the rains bright ochre could hardly be more conspicuous against the green flush of new leaves.

A favourite place for animals to dust-bathe is at the salt-lick. This is often an extensive patch of bare soil where larger animals have mined the earth for the natural mineral deposits. Sodium chloride is not usually present in natural salt-licks; but the soil may be rich in trace elements which concentrate near the surface in arid regions, where there is insufficient rainfall to leach them. Herds of plains game visit the lick irregularly, not necessarily daily as they visit water.

Dust is not a habitat any animal might be expected to live in, yet ant-lions do so. Adults look like damsel flies with knobbed antennae; their larvae are among the few insects that make traps to catch their prey. The ant-lion's trap is a conical

A hen ostrich rising from her dust-bath. Many dry-country birds use dust rather than water to wallow in. The preening and oiling of the plumage that goes on after bathing, whether in water or in dust, helps to maintain the feathers in good order.

pit, in the bottom of which the larva lies hidden. Any insect walking on the edge of the pit starts a miniature landslide and is deposited in the waiting jaws. The larva can survive true desert conditions. Only the overhead sun at midday inactivates it in very hot weather, and it can live in dry dust without food for half a year.

During the dry season dust is an almost living feature of the landscape. It rises in clouds from beneath the feet of game and cattle, and guinea fowl kick up clouds as they scrape about to find food. At the same time there is the wind, not blowing steadily from any one direction, but in sudden strong gusts as air rushes from one local hot spot to another. Some very hairy moth caterpillars and beetle larvae roll themselves up and are bowled along the ground by the wind. A few plants, such as the small succulent *Carallumas* with parachute seeds, rely on distribution by wind; and some acacias are mainly wind-pollinated. Wind tosses the manes and tails of plains game, the crests and plumes of birds, and throws up waves and spume around lake shores. Large animals welcome the breeze for its cooling effect during the hot dry days, and for the dispersal of the many biting, sucking, or simply tickling flies that are otherwise a continual irritation to them.

As the ground heats up during the morning, columns of hot air rise. The wind sucks up the dust

A zebra enjoys a vigorous roll in a dusty depression. While the zebra is in such a vulnerable position a kongoni keeps watch for the possible approach of lions.

Ant-lion larvae live in dry dust, in which they excavate conical pitfalls for ants and other small crawling insects. When it is ready to pupate, a larva spins a spherical cocoon around itself. Sand grains stick to the outside of the cocoon, effectively camouflaging it. After many dry months, rain penetrating the sand stimulates the ant-lion to cut through the cocoon. The adult insect emerges and hangs up to expand and harden its wings, leaving the old pupal skin, papoose-like, looking out of the cocoon.

A pillar of dust whirls briefly across the dry plain, like a live thing. Constant trampling by wildebeest and other game coming to the waterhole has destroyed the grass cover and subjected the soil to wind erosion.

Fire used by the pastoral peoples to clear bush and bring on green grass for cattle, has long been a major ecological force in Africa. Here kites converge on a fire to snap up small animals as they fly or run from the flames.

into spectacular dust-devils that roam the bush, drawing up soil, dead leaves, small insects and any small loose objects in their path. Dust columns may be 60 to 90 metres tall and roar for a kilometre or more across the plain before blowing themselves out. Small animals such as the harvester ants have to stop work in the windy season, as they are very easily blown off their roads by a gust. The mounds of husks accumulated outside their burrows during calm weather are quickly dispersed by the wind and the ants subsist on grass-seeds stored underground.

Much of the heavier material picked up by the winds is soon dropped again, but the finest particles are carried to high altitudes and transported great distances before they fall back on the soil. At the height of the dry season the whole land may be covered by a fine dust. Sometimes the amount in the air is so great that even at midday the sun is only visible as a pale disc through the haze, and an orange light suffuses everything. When smoke particles from numerous grass and forest fires are added to the atmosphere, there is almost an orange fog. At this time of year the sun goes down like a crimson ball, and the sunset colours may be amongst the finest anywhere in the world.

Fire: its effect on the habitat

Fire has been a major ecological force in Africa since time immemorial. It has been the traditional practice of pastoral peoples such as the Masai to set light to the plains during one or both of the dry seasons. This has the effect of keeping down shrubby trees, such as the whistling thorn which is useless to cattle and would otherwise invade the plains; and it maintains the red oat grass and *Pennisetum*, making this sprout fresh growing shoots at a time when fodder for cattle is most scarce.

In some parts of Africa, particularly in hilly areas, annual bush fires have severely damaged and devalued the habitat. But in other parts, burning has produced stable areas covered with nutritious grasses and other perennials that protect the soil and provide excellent fodder. The vast plains and open woodlands which maintain huge herds of plains game—which in turn support the great carnivores and scavengers—have resulted from the prolonged effects of annual fires. Bush

The black-winged plover is attracted to burnt areas after a grass fire. Fresh, green grass blades are already sprouting from among the charred stubble of red-oat and Rhodes grass, and will be eagerly sought by the herbivores.

fires are today as important as topography and climate for the vegetation and fauna of a region.

The main effect of fire is to develop or encourage special fire-tolerant plants and animals which depend on periodic burning. Fire-tolerant trees have thick, corky bark; some produce seeds with a woody integument which explodes in heat, thus hastening germination. Many fire-tolerant plants have large underground bulbs or tubers and slender ephemeral parts above ground. These are known as *geophytes*; they often burst into flower soon after a fire when the ground is otherwise bare and black. Grasses sprout at once after a fire; they are able to survive burning because they have extensive systems of underground rhizomes from which they produce fresh shoots when the old growth is destroyed. New fire-induced shoots of star-grass and red oat grass are a better source of protein, calcium and phosphorus than the taller older growths, so they are preferred not only by domestic cattle but by the wild herbivores, which flock to a recently burned area and may seriously over-graze it. Other animals also flock in, but for different reasons. Marabous, kites, pied crows and others find small dead animals killed by the fire. Living prey is easier to catch with its cover gone, so smaller insectivorous birds also appear. Wheatears, black-winged plovers and Temminck's coursers are often attracted to burned ground.

Many plants have not developed a resistance to burning, especially forest plants which are not normally subjected to fire. Fire can make little headway in forest unless it is assisted by man or, as we shall see, elephants. Forest felling and clearance with shifting cultivation result in the spread of grass where none grew before. This allows fire to enter. Most vulnerable are forest edges and clearings. Once fire has penetrated, withdrawal of the rain forest can be rapid and is almost always irreversible, for fire prevents the regeneration of the forest trees. Forest is replaced by savannah, dotted with small fire-resistant trees. Energy is lost and the soil impoverished each time the vegetation goes up in smoke. With loss of ground cover, erosion is accelerated. With loss of water, climate is affected. In the end, whole regions may turn gradually from forests into deserts.

Vegetation is not the only thing that suffers from fire. With the destruction of their habitat, forest animals disappear and even in fire-tolerant habitats the annual destruction of small animal life must be colossal. Slow terrestrial creatures are the chief sufferers: tortoises and chameleons and giant snails which cannot move fast even to save their lives. In the dry season the bleached shells of giant snails can be seen in places scattered by the hundreds all over the bush, among the silvery skeletons of trees. The young of some ground-nesting birds are also at high risk. Ostrich eggshells are thick enough to withstand fire, and the young are soon large enough and strong enough to escape the flames by running; but the chicks of other ground-nesting birds are not. These species therefore usually nest at the time of lowest risk from fire, either at the beginning of the dry season when the grass is too green to burn, or at the end of the dry season when it is already burnt.

During a fire insects face an additional hazard to the flames, for flocks of opportunist birds are attracted by the easy pickings. Hundreds of kites and large swallows swoop through the smoke to pick off grasshoppers and other insects in the air, while tawny eagles, drongos, black-shouldered kites, marabous and others pounce round the fringe, snapping up rodents, snakes and lizards which are fleeing before the blaze.

It is possible that some rodents and other burrowing animals may escape being burnt if they remain underground. Generally grass fires are not very hot at soil level, although the temperature may sometimes be very high when there is a good layer of dry humus. So the grass nests of vlei rats in clumps of tussock grass are completely charred through, but grass rat burrows remain cool only

94

Two young wildebeest bulls spar with one another, clashing horns and raising a cloud of dust in what appears to be a fierce fight but is in fact a harmless trial of strength.

a few centimetres underground. The rats that survive are probably able to find enough food—roots and fresh-sprouting grass-blades,—but they suffer from lack of cover in which to feed, and so are quickly finished off by predators. More mobile nocturnal species, such as the multi-mammate rat, are prepared to run the gauntlet of the predators and will cross quite large areas of bare ground to reach new living places.

The majority of conservationists are opposed to the use of fire, seeing it as the principal agent in the degradation of habitats over much of Africa. But soil and vegetation vary greatly and there are some regions where the effects of annual burning are beneficial, as in the Kidepo Valley of Uganda. Here the burning is carried out in a controlled way by biologists rather than by herdsmen, to prevent open grassland from reverting to dense bush, which is undesirable both for grazing herbivores and for the people who come to look at them. Not only is the area over which the fire spreads controlled but also the time of year at which the grass is burnt. Burning early in the dry season when the grass is still slightly green produces fires that are less fierce and therefore less destructive and more easily controlled.

Migration

Grasses and grazers, bushes and browsers, evolved at about the same geological time and have balanced each other ever since. When fodder is in short supply, the herbivores move to areas of better grazing or browsing. Migration is the natural device for making best use of land, and the majority of ungulates show some seasonal movement. During the dry season those species that are dependent on their daily drink, such as zebra, wildebeest, buffalo, eland, and Thomson's gazelle, concentrate within a day's walk of permanent water, while those that can do without drinking, chiefly impala and Grant's gazelle, move out of that area and join oryx and gerenuk in the waterless places. Waterbuck may also move: they go not to dry places but to permanent swamps.

The migration of wildebeests and zebras is the most spectacular of the seasonal big game movements in East Africa. From the short-grass plains of the Serengeti, half a million animals set out in May and early June and trek 240 kilometres to their dry-season grazing grounds where there is permanent water and plenty of shade trees. The herds travel in long columns, wending their way over the grasslands. While on the move, the wildebeest bulls maintain movable territories. They stay outside the herds of cows and young and defend their immediate surroundings against other bulls.

On the whole most antelopes are quiet creatures, but wildebeests are rarely silent. The bulls low, bleat, grunt and snort as they cavort about. The younger bulls engage in ritual pushing matches and trials of strength, but rapidly make off in a cloud of dust at the purposeful and lordly appearance of a dominant herd bull. Courtship in wildebeests, as in other antelopes, contains many stylized behaviour patterns. In one display, usually performed at some distance from the herd of cows, the bull rears onto his hind legs, presumably presenting a formidable appearance to his adversaries.

During the dry season impala disperse through the dry, apparently inhospitable bush. The main reason for this is that they feed on the pods of certain acacias which ripen and fall during the dry season. The pods of most acacias are thin and papery, and burst while on the tree. The seeds shower onto the ground, the empty pods remain hanging on the tree. Browsing animals eat the pods, but do not pick up the scattered seeds from the ground below. However, the pods of *Acacia tortilis* do not burst. They are fleshy, large and heavy, and when ripe they fall to the ground without shedding their seeds. These pods have a strong smell which is very attractive to herbivores. Being rich in carbohydrates they form an important part of the dry season diet of impala and kudu, as well as of grass rats, which even put on fat at this season. In over-grazed areas where there is little or no grass left in the dry season cattle also eat the pods.

Impala, then, in some areas are dependent on the pods of this acacia for subsistence during the dry season. In turn, the acacia is dependent on impala for the dispersal and germination of its seeds. When ungulates chew the pods, most of the hard, smooth, rounded seeds are not crushed and are also unharmed by their passage through the animal's gut. Indeed, they do not germinate unless they have passed through the gut of a ruminant. Two more species are closely involved in this cycle: large mound-building termites and star-grass. Like impala, termites are dependent on this acacia for food; in some areas dead thorntrees are their staple diet. The large mounds of the termites are scattered all through *Acacia tortilis* country. On old or inactive termite hills grows star-grass, the grass which is most sought after by impala—who may manure it while eating it. Thus impala spread the acacia, the acacia feeds the impala and the termites; the termites encourage the star-grass, which also feeds the impala; the impala help to nourish the grass.

Every part of the baobab tree is edible. During the dry season elephants attack the trunks, gouging out the fibrous wood. Few trees are left unscarred, and some have been ripped to pieces.

Elephants

Elephants join wildebeests and zebras near permanent water at this time. During the dry season they subsist mainly on the bark and twigs of trees and bushes near rivers or permanent waterholes. They use their tusks to gouge off the tough bark of thorntrees, and whittle twigs between their teeth to remove the bark. They chew wild sisal leaves to extract the moisture, and dig for tubers with tusks and feet. Of all large wild animals, elephants are the most adaptable in their choice of food plants, and feed on any of a hundred different varieties from head-high coarse grasses rejected by other grazers to delicacies such as underground bulbs. Much of what they eat is woody or fibrous, and because their digestive systems are not very efficient an elephant needs to consume about 150 kilos of plant material each day. With such an enormous intake the effect of large herds of elephants on their habitat is profound.

Sometimes the ecological influence of elephants may be beneficial. When they move into an area of swampland, by mowing and trampling they reduce a tangle of two-metre grasses to a freshly-sprouting greensward that other animals can utilize. Or they may sow whole new groves of *Borassus* palms by eating the fruit and depositing the unbreakably hard seeds several miles away. In other places, particularly in wooded country, the effects of elephants are definitely harmful, not only to the vegetation, but to game, including the elephants themselves. In the small isolated patch of high hardwood forest in the Murchison Falls Park, and in many *Terminalia* woodlands, the elephants damage and even kill the trees by browsing and ring-barking them. By trampling the undergrowth they let in light so that grass can now grow. Once grass is established, fires can penetrate the forest and complete the work of destruction. In drier thornbush areas, large concentrations of elephants, with the help of fire, rapidly convert the countryside from bush to grassland. Trees are pushed over and with the large accumulation of dead trunks and branches, and the invasion of the opened areas by grasses, fierce fires sweep through, destroying regeneration. Thus elephants, aided by fire, are the prime natural converters of forest to bush and bush to grassland.

An elephant with her young calf leaves a small seasonal waterhole. Trampling and puddling the wet mud as she drinks, she helps enlarge and maintain the pool, which, as a result, will hold water for longer into the next dry season.

Drought

The destructive influence of elephants is particularly felt in drought years when food is scarcest and competition for it keenest. During the very severe drought in Kenya in 1961 the large herds of elephants in Tsavo National Park were extremely short of food. Very little greenery was left along the Galana river. Herbs and shrubs were eaten down to the ground, and plants which elephants normally never touch were eaten. Mostly the elephants subsisted on the dry bark and leafless twigs of trees and shrubs which, in spite of their unappetizing appearance, maintain a relatively high nutritional value. Destruction was far greater than in a normal dry season, and very heavy browsing and ring-barking killed hundreds of thousands of trees. The elephants also destroyed baobab trees at an unprecedented rate. They have not always eaten baobabs, but pressure of population here and elsewhere has caused them to turn to this new food. In the late 1950s they started attacking baobabs, and now few large trees remain unscarred, while many are in process of being spectacularly destroyed. All parts of a baobab are edible, so that not only are branches, bark, leaves, flowers and seeds eaten, but whole large trees, which may be a thousand years old, are ripped apart for the moisture and calcium content of the fibrous wood. Fire completes the destruction by killing the seedlings. Almost as if in retribution, there were several instances of elephants being killed by the falling trunks of the baobabs they were eating.

It is not known how browsing antelopes such as kudu and gerenuk suffered in competition from the elephants during the 1961 drought. Being nomadic to some extent, they were probably able to move out to less heavily browsed areas which the elephants could not use because of lack of water. But the black rhinoceroses suffered very severely, to the point of starvation. Along a 64 kilometre stretch of the Athi river alone, 282 rhinoceroses starved to death, though it is not certain that the elephants were wholly to blame. Black rhinoceroses never move from their home range in the dry season, even when competition from elephants is severe. If their territory does not contain permanent water, they do without rather than make long treks to it as elephants do. Rhinoceroses are particularly selective over their food in the dry season, and will eat only the green parts of plants in preference to dry withered bits, to obtain the moisture they need.

During the dry season the black rhinoceros selects green leaves and twigs wherever it can, but during severe drought it must subsist on dry twigs and bark and such moisture-rich though bitter succulents as are available on its home range.

Grazing is almost gone in this drought stricken area, where rain has not fallen for almost a year; but this herd of eland does not look distressed. They can subsist quite well on the protein-rich twigs of acacias. The prickly pear cactus is an introduced weed, but in districts like this may provide fodder for animals otherwise desperately short of food.

The Kei apple is another introduced hedging plant benefiting many animals during lean periods when indigenous food is scarce. Long-tailed thicket rats live on the apples and nest in the dense thickets which this small tree provides.

They also chew succulent latex-bearing plants, wild sisal and euphorbia. Stems or leaves may be chewed for as long as twenty minutes to extract the moisture, after which the fibres are spat out. Finger euphorbia is eaten in great quantities where it is common, and much sought after where it is less common. To get at the higher branches, a rhinoceros stands up on its hind legs and climbs with its front feet as high as it can. It then hooks its horn among the branches, and walking backwards, snaps off large amounts of tree. It eats all the small branches and gouges the bark off the bigger ones with its front horn and teeth.

Finger euphorbia is not indigenous to East Africa, but was introduced from India. Its colonization and spread throughout the region must have had a profound influence on the survival of the rhinoceros in arid areas without water. Another introduced succulent, the prickly pear cactus, of which four species have been brought from South America, may also have influenced the survival of some animals in very dry areas. The thornless type is a valuable dry-season fodder for cattle, but the thorny type is more widespread as it has been much used as a hedging plant. In a few years, it develops into dense thickets, which provide cover for many kinds of small animals. Its spines deter creatures from eating it except in time of direst need, when it is chewed by the larger ungulates. However, it is of greatest value to smaller creatures

With their grazing destroyed by the domestic herds of the Samburu, these Burchell's zebras are starving. They do not, however, appear skeletal like the cattle: their contour-striping and quite thick, rough coats obscure their thinness.

and may provide their only means of subsistence in a prolonged dry season. During a severe drought grass rats and small birds were seen to be eating not the cactus itself, but the cochineal insect that was introduced in attempts to control the cactus: pink droppings all around the cacti showed what they had been feeding on.

Other introduced hedge plants may also provide locally abundant food during a lean period. The Kei apple produces a bountiful crop of delicious-looking small, yellow apples which ripen and fall in the dry season. During the day the fruits attract flocks of glossy starlings, mousebirds, coqui franco-lins, and brimstone canaries, whose yellow breasts match the yellow apples; while at night they are eaten by small game such as steinbuck and thicket rats.

In a severe drought animals are often forced to make do with unpalatable or unusual alternatives to their diet. Warthogs dig for roots and tubers with their tushes when grass is scarce; rock hyraxes chew the stems of the large *Caralluma* which even goats do not touch; bushbabies and pottos eat tree gum. Even lions may chew the stems of wild sisal to obtain moisture, although in a normal dry season they get enough water from the stomach contents

of their prey. Cheetahs, too, lap the blood that collects in the rib-cage of their kill as in a bowl, and obtain sufficient moisture that way.

In a normal dry season the wild animals of Africa do not suffer hunger or thirst. Over millenia they have adapted to heat and periodic drought. The animals that really suffer are the cattle, especially during drought years. Cattle need to drink at least once a day, which means there is a limit to the distance they can graze from water. When many waterholes dry up, all the cattle herd around the few that remain, grazing out all the grass for miles around. Other grazing animals suffer from the concentration of cattle. Worst hit are the zebras, which under normal conditions come through a drought remarkably well, since they can digest coarse dried grass which is useless to other animals. During the 1971 drought, however, zebras in the Maralal Game Reserve were

A zebra carcass, only half cleaned by vultures, is left to dry in the sun and wind. Perhaps the grass will grow more richly around this corpse when the rains finally come.

Yearling zebras fare worst in a drought. Jackals have eaten the hindquarters of this carcass, leaving the rest as a feast for Egyptian vultures.

dying in hundreds. Yearlings suffered most; older beasts and foals at foot fared better. The Samburu cattle were also dying, for there simply was no grazing. Only the goats looked sleek and fat, and it was a sad sight to see hordes of them fanned out over the hillsides, nibbling down to the bare earth what little stubble remained. They also ate the protein-rich twigs of thornbushes denied to cattle and zebras, but fortunately available above goat level for the taller wild herbivores, giraffe, eland, impala and gazelles.

The drought of 1961 is said to have killed three-fifths of the cattle in Kenya, and the drought of 1971 also caused severe losses. The problem of survival in a drought is made worse when, as often happens, it is preceded by several years of good rains. Then the grass is plentiful and the cattle increase. To supply the big herds with water, new wells are dug. But every year the grazing shrinks. A larger number of big animals can quench their thirst, but they cannot find food, and the pressure on the vegetation destroys it further. In places, more than ten times the number of cattle are maintained than the country can support in times of drought. In some areas the proliferation of artificial wells has diverted water from what used

to be permanent waterholes for wild animals. The result has been a reduction in game because, over wide areas, the amount and distribution of permanent water is the limiting factor determining the number and distribution of game.

During a severe drought, there is a feeling of suspended animation in the parched plains and the grey bush, an air of simply existing if possible, rather than living. As pools dry up and turn to pans of cracked mud, crocodiles lie lethargically waiting for the water to return. Insects and other invertebrate life seem to have died out, yet flocks of starlings and troops of baboons are finding something somewhere to sustain them—perhaps a few termites under still-damp elephant dung or a solitary wolf spider waiting in its burrow. Only the carnivores and scavengers have plenty to eat as carcases of dead zebras and cattle litter the plains. Many are left to dry in the sun: they are too numerous for the scavengers to dispose of. Throughout each long, hard, hot, dry day there is a feeling that all animal life is simply enduring and waiting; waiting for clouds to begin gathering, for humidity and softness to return to the air and the earth, for the rains that will make the scorched and barren countryside spring to life once again.

The Wet Season

On the equator, rain falls in two seasons, just after the equinoxes in April and November. Throughout East Africa these wet seasons are modified by distance from the equator and from the sea, and by altitude. No part receives rain evenly distributed throughout the year; one of the two wet seasons is longer and wetter than the other. In many places the total rainfall is less than 50 centimetres a year, grading to as little as 25 centimetres in the northern deserts. In these places the rainy seasons are separated by very long droughts. When the rain falls, there is sufficient only for a brief flush of annual herbs and grasses, and a short intensive breeding season for birds and other animals. In other places there may be 150 to 175 centimetres a year and in the mountains much more, with hardly a month in which there is no rain. The vegetation of the higher rainfall areas used to be forest, but today much of it is wooded savannah, even grassland. Here the differences between the seasons are not dramatic. Flowers bloom the year round, vege-

tation remains green and birds nest at any season. But in the lower rainfall areas the change is very dramatic, from parched inactivity in the dry season to intense activity, in the rains.

Rain falls in the tropics, as elsewhere, when winds converge. Air rises above an area of convergence and cools by expansion so that moisture in it condenses, then precipitates as rain. Two main convergence zones affect East Africa. The largest and most important is the equatorial trough or *intertropical convergence zone* (ITCZ), a belt where trade winds or monsoons from the southern and northern hemispheres meet. This zone brings rain as it moves seasonally northwards and southwards following the sun, but is much modified by smaller, more complex, local wind patterns. The ITCZ affects much of East Africa, but rain is also brought to the Western Rift by a second semi-permanent convergence zone, the *Congo air boundary*, where Atlantic westerlies meet easterlies from the Indian Ocean.

Previous page: A young gerenuk buck among the spring-like greenery of fresh leaves and grasses at the beginning of the rains. The first showers of the wet season fell only days before, transforming the dusty, grey bush almost overnight into an emerald paradise.

Below: The first storm of the season, coming across the lake in the later afternoon, whips the murky water into wavelets. The flamingos become very excited, going through the motions of bathing in the downpour and taking off across the lake, apparently beneath the very arch of the rainbow.

Rainwater floods down into the soda lakes from the surrounding hills. Lake birds such as black-winged stilts and a ringed plover gather at the river mouths to bathe excitedly in the pools of water.

Right: Even before any rain has fallen the increased humidity of the coming rains stimulates acacias to blossom, providing copious nectar for insects.

Below right: Birds become very excited during the first storms. A female bronzy sunbird fluffs out her feathers and flutters her wings as she enjoys the shower bath.

Towards the end of the long dry season clouds begin to build up daily as the atmosphere becomes perceptibly humid. In response, every acacia bush bursts into blossom and leaf, filling the oppressive atmosphere with refreshing scent and providing welcome fresh browse for herbivores. The acacias' early flowering allows wind or insects to pollinate them before the first heavy showers batter the flowers to pieces. Now bright colours return to the bush. The flush of new leaves in acacia bushland is emerald, but other trees flush red. The *Brachystegia* woodlands of southern Tanzania break first into spectacular bright red or copper leaf, then gradually become duller, through olive to their normal glossy bright green.

When the long-awaited rain comes, it does not fall as a gentle drizzle but typically as a sudden

After the main downpour this speckled pigeon came out from shelter and just stood out in the rain, allowing the drops to settle on its plumage. Afterwards it preened the more vigorously for having got its feathers damp.

In arid areas where there is little grass left at the end of the dry season, tiny fragile-seeming clusters of white lilies spring up out of the hard ground as soon as the first rains have fallen. Small beetles, such as this longicorn, and tiny chafer are attracted to the flowers and soon spoil them by eating the petals.

The desert rose looks like a small tree; it grows in arid regions and stores moisture in its bulbous lower trunks. Its pink blossoms are particularly lovely against the new greenery of sprouting acacia.

heavy downpour, often in the late afternoon or just after dark. Everywhere there is great excitement at the coming of the rains. Even when the fall is many miles away zebra families congregate into large herds and the stallions run about braying excitedly before they all set off towards the storm. They have been seen galloping with wildebeests towards a storm 8 kilometres away, and will travel 40 kilometres in a night to reach a more distant one. Grant's and Thomson's gazelles will cover 15 kilometres in a night to reach the site of a first rainfall. During the rain great grown lions may splash and gambol in it like cubs. Wild dogs become very excited and run round licking the rain from each other's coats. Birds welcome it as a chance of a showerbath. Hornbills and sunbirds particularly, which do not normally bathe in standing water, flutter with great excitement in the downpour and among the rain-soaked vegetation. Parrots are specialist rain-bathers, and have characteristic spread-eagled and fluffed out postures for rain-bathing. Pigeons and doves rain-bathe in the same posture as they sun-bathe, lying over on their side with one wing extended and raised to let the rain get to their flank and under-wing. Larks lie out in the rain with their wings

Caralluimas, found only in semi-desert, are the African equivalent of small cacti. At the beginning of the rains their stems swell again with moisture, and buds appear. The large five-petalled waxy blossoms burst open suddenly with a distinct little "pop". Like much other succulent greenery at this time of year, stems and flowers are soon smothered in rapidly-reproducing aphids. The tiny gerbil is holding its hind foot while it washes its toes.

Above right: The beautiful, large-flowered Crinum lily appears in grasslands soon after the first rains, its flowers opening in succession and quickly dropping. The mushroom-like agaric may indicate a termites' underground fungus garden.

spread. Even normally water-bathing birds become very excited during the showers and make clumsy attempts to rain-bathe.

As soon as the first rains have fallen in the bush many plants burst spectacularly into blossom. Plants with large underground bulbs or tubers flower within days of the first rain: delicate white lilies push up through the hard bare earth, and the lovely deep pink blossoms of the desert rose open. Aloes send up tall spikes of orange trumpets, providing an abundance of nectar for sunbirds and bees. Small succulent caralluimas that have escaped the goats in thornscrub thickets put out surprisingly large, starlike waxy blossoms, and giant euphorbias are covered with tiny, sulphur-yellow flowers.

The most surprising plant growths in the bush are the fungi, which force their way up out of the earth a few days after rain. Often toadstools appear in "fairy rings" around termite towers. They are the *agarics* or fruiting bodies from fungus gardens cultivated underground by the termites. Worker termites make elaborate sponge-like combs out of undigested cellulose on which their special fungus grows. They do not eat the fungus itself, except for the small white fruiting bodies or *conidia* which may be a valuable source of vitamins; but by providing warmth and moisture the gardens play an important part in air-conditioning the underground nests. The toadstools are a second form of fruiting body, but only appear seasonally. They may spring up anywhere in the bush as well as around termitaria, and locate otherwise unlandmarked termites' nests beneath the soil.

On the plains, too, many plants come into flower and the grass begins to sprout, transforming the blackened or straw-coloured land. Elephants reverse their dry-season movements and wander away from permanent water, spreading out through the bush where they rely on numerous small wet-season waterholes. They now eat vast quantities of fresh green grass, not indiscriminately, but selecting certain species and ignoring others, plucking the upper shoots and discarding the roots and stalks. They become very excited and push over trees, particularly *Commiphora*, without feeding on them. This strange behaviour may simply be playful trials of strength when they are in a general state of excitement.

Baby mammals

With the abundance of good grazing, most other ungulates move out from the permanent-water areas where they had concentrated during the dry season and disperse widely over the countryside. At the same time impala, waterbuck and Grant's gazelles, which had dispersed in the dry season, move back in large numbers. Many give birth now. Smaller antelopes have two breeding cycles a year, but most of the larger plains game have one annual breeding cycle and drop their young at the

start of the long rains, when feeding conditions are at their best.

With the coming of the rains the great herds of wildebeest migrate back to the plains to calve. The way is led by Thomson's gazelles and zebras, which may be more sensitive to barometric pressure or to wind and temperature changes, and so are aware when distant rain has fallen. The long columns of wildebeest follow. The wildebeest return to their regular calving ground, for instance in the Ngorongoro Crater and the Loita plains, when these are sprouting with fresh green grasses. The herds follow well-defined traditional tracks making deep gullies across ridges. On the calving grounds all the cows give birth within a short period of a few weeks. Parturition takes place in the open and is usually very quick. The calf may be on its feet within minutes of birth, and will at once try to suckle, although perhaps from the wrong corner of its dam. Within only five minutes of its first wobbly efforts at standing it can run with the nursery herd. Once on its feet and moving it is able to keep going and keep up with its mother. Jackals may wean their cubs on the wildebeests' afterbirths, so plentiful are they at this time.

Other plains game, such as hartebeest and zebra, may also calve when grazing is best, so that they coincide with the great wildebeest birth peak. Most antelopes are born into nursery herds, but zebras are born into family groups of a stallion, his mares and their last year's young. The stallion will actively try to defend his family from hyenas and wild dogs. Many antelope and gazelle mothers are also exceedingly brave in defence of their young. Oryx are a match even for lion, and a Thomson's gazelle will charge the smaller predators—jackals, baboons, martial eagles or secretary birds—sometimes succeeding in driving them away. At birth a gazelle fawn's coat is a much darker brown than that of the mother, and it will try to escape detection by lying motionless close to the ground.

The camouflage of baby antelopes is probably coincidental. Many lack the disruptive white spots of deer fawns, but their colours blend effectively with their surroundings. The brown baby coat of the gazelle fawn, for instance, blends well with old or new grass. But even so, during the breeding season a very heavy toll of fawns is taken by leopards, cheetahs, wild dogs, hyenas and jackals; and occasionally also by lions, servals, caracals and the larger birds of prey.

In the same spring-like season, vervet and patas monkeys and baboon babies are also being born At birth, the ears of baboon babies are oversize, crinkly and pink. Their faces are pink, their muzzles corrugated and foreheads puckered. Pink faces and very dark body fur arouse strong emotional reactions in adult baboons, so that babies become the centre of interest for the whole troop, mothered by all the females and fiercely defended by the adult males. Like antelopes, monkeys normally give birth to only one young at a time. The need of antelopes to produce young fully active at birth precludes the development of more than one at a time, or very occasionally twins. With monkeys, it is an adaptation to life in the trees. The mother must be able to move about freely, even while carrying a new-born baby, and the baby must be born able to cling tightly to her, to leave her hands free for climbing and feeding.

A baby steinbuck is left hidden down an old ant-bear burrow or in the grass, with which it blends perfectly. Steinbuck and other small antelope fawns may be born at any time of the year, since they have two annual breeding cycles. Even so, a high percentage of them are likely to be born now when conditions are best.

Baby vervet monkeys are also most likely to be born at the beginning of the rains. At first the baby clings to his mother's belly, and peeps out upside down between her front legs when she is on the move. When she sits down to feed or rest he is already in the most convenient position for suckling.

Left: Kongoni or Coke's hartebeest are most likely to produce their young while the grazing is best, early in the rains.

A baboon baby at first clings upside down under his mother's belly, but by the time he is four months old he can ride jockey on her with confidence.

Zebra mice babies in their grass nest. They are born completely helpless, their eyes and ears closed, with a faint covering of downy fur. They will need to be kept warm by the mother for two weeks before they are ready for short excursions.

Elephants become particularly excited during the first storms of the rainy season, and go around pushing over small trees— apparently just for fun.

Spiny mice are unique among rats and mice in giving birth to precocious young. This baby, only one day old, is well furred though not yet prickly; and its ears and eyes are open.

Beisa oryx on the sub-desert steppes of northern Kenya. Only days before, these plains were parched and brown; the first rains coincided with this calf's birth and produced a flush of good grazing for the lactating mother.

A young monkey's age can be told by the position in which it clings to its mother. At first it clings beneath the belly, with arms around her and fingers and toes tightly locked in her fur. It peeps out, upside down, between her front legs. A few weeks later it will try to ride on her back, but cannot yet sit upright. Not until it is about four months old can it ride jockey with confidence.

Like monkeys, bats carry their one baby clinging to their fur. But single births are unusual among mammals as a whole; multiple births are the general rule. The record for litter size among African mammals is held by the multi-mammate rat, which may give birth to any number up to twenty at a time. The female has a very large number of teats, up to twelve pairs, whereas other rats and mice normally have no more than five pairs. The babies are born blind and naked, but develop rapidly until by the time they are twenty days old they can fend for themselves. The mother can give birth every twenty-five days, while her babies can reproduce at two months old, so the multi-mammate population can expand five-fold every month during a good rainy season. In exceptionally wet years, when the breeding season is prolonged, the population may build up to

plague proportions. Grass rats may also build up their numbers at the same time, and can be seen out in the daytime feeding on the new grass and scampering across open places, while multi-mammate rats come out at night to feed, and may be heard rustling and sqeaking everywhere.

Litter size in mammals is usually inversely related to the stage of development of the young at birth. At the other end of the scale from the multi-mammate rat comes the spiny mouse, which is unique among rats and mice in giving birth to fully active, precocious young. A female spiny mouse has a very long gestation period for her size, about thirty-eight days. The two or three precocious babies in a litter have their eyes open at birth, and can totter around the nest area straight away. They are well furred, but not yet prickly. In a few days they are nimble and active, at a fortnight weaned, and sexually mature, ready to reproduce, when they are about seven weeks.

Flash floods

As the wet season progresses, rainstorms become more frequent and are often very violent, with lightning and thunder. Animals are no longer excited by the rain. Birds of open country caught by a heavy downpour adopt a special rain posture, standing upright with the feathers sleeked against the body so that the rain runs off. Large animals shelter under trees, standing close together. Plains game stand with their rumps towards the worst of the storm, their backs glistening characteristically. At high altitudes, even on the equator, these very violent rains sometimes fall as hail, which shreds

111

vegetation and does much damage in agricultural zones. When the storm has passed, it may leave whole areas of grassland flooded, so that for a time the problem is one of too much water where previously there had been too little. Herons, ibises and storks wade about feeding on the small animals flooded from their burrows.

In grasslands, the permanent plant cover provides good protection for the soil during violent storms and floods. Forests, with their tall trees and undergrowth and complex root systems are even better protectors of the soil. But where land has been denuded of vegetation by over-grazing or cultivation, violent rains are a powerful agent of soil destruction. Only part of the downpour penetrates the soil, the rest runs off downhill, forming small torrents which carry away humus and soil. In hilly country, the run-off scores deep parallel gullies; at every rainfall the gullies deepen until in a few years they may have eroded into seventeen-metre ravines down which the flash floods roar, carrying with them surface soil that may have taken natural vegetation thousands of years to lay down. On gentler slopes, sheet erosion is less spectacular, since at first it shows only as a change in the colour of the soil. Gradually, as the finer particles are removed, a layer of pebbles is left at the surface. Bare soil lacks the power to absorb and hold water, and evaporation from it increases. In this poor leached soil, baked hard and sterile by the sun, only a few desert-adapted plants can grow.

One of the adaptations to an arid environment which enables acacia trees to flourish in semi-desert is the development of very long root systems which can penetrate deep into the soil to tap moisture unavailable to shallower-rooted plants. After a heavy shower acacia seeds that have been lying on the bare earth beneath the parent tree get swept up and carried along by the run-off. The soaking causes them to swell and germinate. The aerial parts of the seedlings develop slowly, but the roots grow much faster. However, the moisture does not penetrate far into the ground, and it quickly evaporates, so that the seedlings will find their roots trying to penetrate waterless earth and most will die before the next rainy season. It is not clear how the seedlings that survive are able to do so, or how they can grow through moistureless soil to attain the root-lengths of the adult plants. The problem regarding drought-resistant trees is not how the grown trees survive, but how the seedlings ever reach maturity.

The northern Uaso Nyiro River, northern Kenya, in spate after rain, carries along with it tons of surface soil that has been eroded from the surrounding land during the violent downpours of the last few days. Only a week ago, the river bed was almost completely dry and the crocodiles were starving. In another day, as its level falls a little, the crocodiles will find plenty of corpses of animals left by the flood.

Acacia seeds germinating after rain. Their green parts will grow slowly, but their roots push down fast to find underground water before the soil dries out again.

112

Much of the rain that falls on the thirsty bush country scarcely benefits that countryside at all, for only a small proportion of it soaks into the ground. The rest runs off in rills and gullies into temporary water courses. These disgorge into the big rivers which become raging torrents, bearing along trees, dead animals and silt. Under the pressure banks collapse, especially where in the dry season elephants have dug beneath them in the dry river bed to find water in the sand. Several of the big rivers, for example the two Uaso Nyiros in Kenya, lose themselves in vast swamps, while others drain into lakes which may themselves have no visible outlet. Other rivers have their outlet in the sea. So by evaporation or run-off most water is lost to the soil.

When a very heavy downpour occurs in the region of Lake Magadi or Nakuru a spectacular die-off of tilapia may result. Probably an algal flush follows the influx of fresh water, and this leads to deoxygenation of the lake water and suffocation of the fish. At Lake Nakuru half a million dead fish were once washed up around the shores after a rainstorm. Simultaneously all the fishing birds departed from the lake, and the flamingo hordes left overnight. The lake then had a most strange appearance, almost deserted of birds except for dabchicks and a few egrets, and an exceptional number of marabous, pied crows, kites and other avian scavengers which descended on the shore to feed on the sun-baked fish. But half a million tilapia are, after all, only a week's food supply for all the fish-eating Nakuru birds. The surviving fish soon built up their numbers and pelicans, flamingos and all the other birds returned to the lake in as great multitudes as before.

Animals of the rainy season

During the rains the sky is overcast, with low oppressive banks of grey cumulus. Showers may fall all around, even if they do not hit the same area again for some weeks. The atmosphere remains humid, and the sun's evaporating power tempered by cloud. Rain means the start of a new life for a large variety of animals that need a damp atmo-

Above: Lake Nakuru, strangely deserted of its flamingos and fishing birds after a rainstorm had caused a spectacular fish die-off there. Little egrets were one of the few residents that remained, while kites flocked in by the score to feast on the dead tilapia.

Giant red velvet mites can survive completely arid conditions for long periods. At the start of the rains myriads of them reappear for a brief, intensive feeding and breeding period.

Millipedes are normally nocturnal, but at the beginning of the rains they may appear by the thousand above ground in the daytime, looking for mates. After pairing they disperse underground again and revert to their nocturnal habits.

Giant snails can remain dormant for well over a year, cemented into their shells by a cap of dried mucus. When rain softens the cap they emerge, to munch green leaves.

sphere in which to carry on normal activity. At the beginning of the previous dry season these animals burrowed into the ground before it became too hard, and remained there throughout the hot dry weather (page 80). Now the rain softens the ground, awakes them from aestivation, and they dig their way out for a short period of intensive feeding and breeding.

One of the most conspicuous invertebrates to appear after rain is the centimetre long velvet mite or "red spider". They creep over the ground, often in myriads in really arid country, their scarlet colour a warning that they are distasteful. Birds, big spiders, even the voracious scorpions and solifugids leave them alone. While the ground is still damp they are diurnal, but as the atmosphere grows less humid they become crepuscular. They mate and lay their eggs within a few days. The young grow very rapidly, feeding on vegetable detritus, then go into aestivation, perhaps for a whole year or more, until the next rains.

Other normally nocturnal creatures that suddenly appear everywhere during the day are very large woodlice. These abound in rather shaded places such as among the leaf litter in riverine or gallery forest. They creep about in broad daylight

for a few days, looking for mates, then revert to their normal nocturnal habits. Woodlice are crustaceans, an order whose other members live underwater or in very damp places. Lacking an efficient waterproof skin they are particularly susceptible to desiccation, so only come above ground by day after very heavy rain.

Multitudes of millipedes appear in the bushlands at this time. There are many sizes. The giants are splendid cylindrical creatures about 120 millimetres long, exquisitely precision-built, moving smoothly forward over the ground with synchronized waves of their 228 legs. All millipedes trundle about seemingly unaware of anything beyond their own antennae. Their assured behaviour suggests that they too, are distasteful, and when alarmed they spring into a coil and exude an astringent substance. But millipedes are a favourite food of the banded mongoose, whose occupation of a termite mound can be diagnosed by the presence of quantities of millipede rings in the mongooses' latrines. Scorpions too, eat millipedes and in spring-cleaning their burrows bulldoze out mounds of millipede rings and earth which they tamp down outside. When a millipede walks by the entrance to a scorpion lair it causes a landslide which deposits it in the scorpion's underground dining-room.

Millipedes may sometimes occur in very great abundance. One species, a glossy black animal with orange legs, sometimes swarms twenty or thirty to the square metre. Some individuals may be feeding, scavenging on decaying plant material, but most of them are proceeding purposefully about looking for a mate. They are so thick on the ground one cannot avoid treading on some; elephants leave a trail of crushed bodies, and the sharp hooves of antelopes cut many in half: they still continue walking without their tail ends. A few birds eat millipedes—fiscal shrikes, red-billed hornbills and cuckoos particularly. After three days the millipedes disappear, not eaten but gone to ground, under logs and stones and elephant dung.

Rain puddle fauna

During the rainy season the downpours fill up all those seasonal waterholes that have remained dust bowls through the drought. Water is now available over a wide area of bush for all the big game, though even those ungulates that normally drink daily require little water during the peak of the rains because of the high moisture content of the lush greenery they eat. When they do visit the waterholes their trampling maintains and enlarges the pools, so that they hold the water for longer into the next dry season. The effect of animal maintained pools on the ecology is very great, since they enable animals to stay in the area well into the dry season. These pools also provide food and drink for butterflies, which suck up moisture and mineral salts from the wet mud where big animals have urinated; and they provide breeding grounds for a host of smaller creatures, insects and other invertebrates, amphibians and fish.

Some seasonal puddles, dry for half the year, miraculously swarm with little fishes during the rains. These are killifishes, amongst the most brightly coloured of all fresh-water fish. They grow to a length of only 40–50 millimetres and the males in breeding colours are living jewels. One species is olive green spotted with pale blue, red and white, with an all-red tail fin. The female is a drab grey-brown. Another is a lovely deep violet, spotted with brilliant red, with yellow pectoral and ventral fins. They live in rainwater puddles and seasonal pools and are excellent jumpers, flipping across the mud to find deeper water when their own pool dries out. When all the pools are dry the killifishes die. Their species is perpetuated by drought-resistant eggs.

Killifish eggs pass the dry season in the dried mud of what was once a pool. When the weather breaks and the puddle fills, most of the eggs begin to develop and hatch within hours. A few eggs do not develop with the first wetting, a safeguard should the rains fail and the pond dry up again too soon. The fry grow extremely fast, feeding at first on the tiny crustacea which also hatch from drought-resistant eggs. As the fishes grow they graduate to larger crustacea and mosquito larvae. In only six to eight weeks they are sexually mature and start breeding. The males are extremely pugnacious, fighting one another and chasing the females. When a ripe female is ready to spawn she allows the male to swim alongside and wrap his dorsal fin over hers. Together they plunge into the mud, disappearing in a cloud of silt. They continue to spawn throughout the short breeding period until the pool dries up and they die. The eggs cannot hatch in the same water in which they were laid. They must go through a dry period of many weeks—twelve is the optimum, but they will readily survive much longer—so there is no danger of the fry hatching out and dying prematurely with their parents.

Temporary puddles have an abundant plankton of small crustaceans, known as "water fleas", which often occur in such millions that the water is soupy with them. Slightly larger crustaceans, like little oval bivalve shellfish, up to 3 millimetres long, are also sometimes abundant in a pool. These are clam-shrimps. They look like tiny translucent molluscs, for their shells completely enclose the body and limbs; but antennae and legs can be extended between the valves to beat and kick the animal through the water. They swim in little jerks like water fleas, scavenging anything microscopic, plant or animal, in the water.

A much larger kind of crustacean that flourishes in puddles is the fairy shrimp, a very primitive animal. Like the lungfish (page 80), it is a living fossil, being similar to very ancient fossil crustacea. Adults are about 2 centimetres long and swim on their backs like the related brine shrimps. They, too, feed by filtering microscopic particles from the water. They are quite beautifully coloured, mostly transparent but with some red on them, and with iridescent greens and blues on the body. The eggs are carried by the females in large egg-sacs. Eggs that have not been dried out may hatch, so that several crops of fairy shrimps can be found in the same puddle. If the puddle dries, however, the eggs can remain viable in the dust for years, and dried eggs

Tree frog tadpoles feed at first by rasping algae; later they become carnivorous and feed on dead fairy shrimps, other tadpoles and so on. The water tiger, larva of a big diving beetle, hatches as a voracious carnivore and remains so throughout its life.

The male of the tiny East African killifish *Nothobranchius guentheri* is among the most colourful of freshwater fishes; the female is more drab. They live in seasonal rainwater pools. When the pool dries out individuals of a season die, but their eggs survive the drought and hatch in the following rains.

actually hatch quicker than those that have not been dried.

Fairy shrimps and water fleas are very important in the ecology of the rain puddle and form a major item in the diet of the next class of puddle fauna, the carnivores. These aestivate through the drought and appear suddenly as soon as rain has fallen. One of the most conspicuous is a large water scorpion, a big, wicked-looking brown bug, with a long slender respiratory snorkel-tube at the end of the abdomen, and mantis-like front legs. It normally flies by night but at the start of the rains it appears all over the bush by day as well, scurrying along in search of a puddle. Water beetles also appear; they are clumsy on land since their legs are oar-like paddles adapted for swimming. Oval in shape and beautifully streamlined, with smoothly polished dark green elytra, these beetles cut through the water, preying on any animal they can overpower. Their exclusively carnivorous larvae are known as water tigers, long, slender creatures with hollow, sickle-shaped mandibles for seizing and sucking their prey.

One of the main prey of water bugs and beetles is tadpoles, and there are countless numbers of them in rain puddles and temporary waterholes. There are no tailed amphibia in East Africa, but a great

Frogs and toads may have one large vocal sac beneath the throat or, as in this big groove-crowned bullfrog, two smaller lateral ones. Toads and reed frogs keep their balloons pumped up all the time, whether they are producing noise or resting between calls. The bullfrog's two small balloons only appear at the actual moment of sound production.

Yellow-billed storks, spoonbills and great white egrets excitedly fish for spawning toads at sunset. The noisy calls of frogs and toads attract not only males and females of their own species but many amphibian-eating birds also, so spawning usually takes place at night when the birds are roosting.

variety of frogs and toads manage to live in almost all habitats except waterless deserts and above the snow-line. Some live in permanently humid environments, in swamps, reedbeds or even hopping about on the forest floor. Others survive in semi-arid areas where there may be no rain for three-quarters of the year. When rain comes, every pond or puddle becomes alive with frogs or toads. They consume between them an astronomical number of insects, and are themselves consumed in vast numbers, especially in the tadpole stage, by all predatory aquatic insects, crabs, wading birds of every kind, kingfishers, snakes, terrapins and other frogs and toads.

The courtship of frogs and toads is usually a noisy affair. As soon as a male finds water he sits in it and "sings". Common toads croak, a deep, loud "Waaa-waaa" which attracts all others of that species within a wide radius. The chorus and splashing attract storks too, so spawning takes place mainly at night when the storks are roosting. A shower of rain stimulates a fine chorus from the tiny reedfrogs, too. Only 2 centimetres long, they are capable of producing astonishingly loud sounds for their size, and when hundreds of them are calling in the reeds their metallic tinkling chorus fills the air. Another characteristic frog noise of the rainy season is a melodious, watery "Boink", loud and penetrating from near to. The call is ventriloquial and seems to come from all over the place, from the water, from grass and bushes, even from up small trees. Any sudden noise stimulates a volley of "boinking". The sounds travel in bursts, and the makers are almost impossible to find. But the running frog, the maker of these calls, is common all over East Africa in open grassland from sea shore almost to snow-line. As its name implies, it seldom jumps, but runs or walks.

Frogs and toads lay large numbers of eggs, in some cases thousands, indicating that mortality is high. Many eggs must hatch in order for one or two young to reach maturity. A pair of common toads, for instance, may lay 24,000 eggs at a time, in strings of protective jelly. Grey tree frogs protect their eggs by laying them inside a foam nest. The female produces a thick mucus-like liquid, which she and her mate whip into a white froth with their hind legs. About 150 eggs are laid in this froth, which is always placed in vegetation over a waterhole. The outer layer of the froth hardens in the sun into a meringue-like crust. In three or four days the eggs hatch, the lower layers of froth liquefy as the larvae swim about, and finally the tadpoles drop into the water.

In contrast to most frogs and toads the courtship of the entirely aquatic clawed frog is an exceptionally quiet affair. The only call it makes is a faint ratchet-like chirping underwater. Its tadpole in the legless stage looks like a small catfish because of the long slender barbel on each side of its wide mouth. Unlike other tadpoles it does not rasp algae but at first sucks in protozoan-rich mud and filters the excess water through gill-like slits on the sides of its head. The adult frog and older larvae have insatiable appetites, which make them of considerable ecological importance. They feed on any animal small enough to be raked into the mouth with their long sensitive fingers. The frogs are well camouflaged by their mottled colour, and their smooth skin makes them difficult to hold; but

The grey tree frog spawns in vegetation above a rain pool, in clumps of grass or on a log, or high in the branches of an overhanging tree. As the eggs are laid, the pair beat the albumen into a froth with the hind legs to form a foam nest.

Inside the froth nest, the grey tree frog's 150 white eggs hatch into tadpoles in 3–4 days. Two days later the tadpoles' wriggling liquefies the froth and softens the crust, and they drop into the water to complete their development in the usual way.

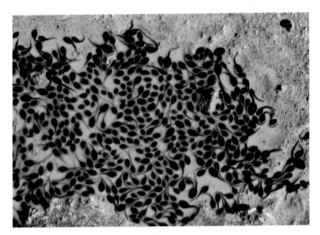

The striped pyxie toad lays its eggs in rain puddles. In some districts where the soil is very porous, the puddles may seep away and evaporate quickly. In shallow water the tadpoles gather in bands.

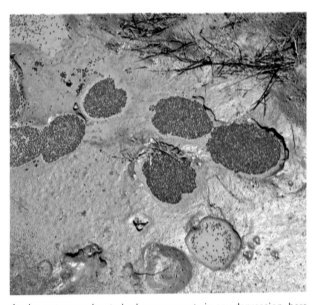

As the water recedes, tadpoles congregate in any depression, here in the hoof-prints of the antelope that drank up much of their already small puddle.

Even small footprints of wader and steinbuck become filled with tadpoles as the puddle evaporates faster than they can develop.

they are preyed upon by large fishes and herons in the lakes and by terrapins, kingfishers and storks in the puddles.

Frogs and toads are most abundant in the areas of higher rainfall where pools are likely to remain for at least three months, long enough for tadpoles to develop at a normal slow but steady rate. However, in certain areas of less reliable rainfall, particularly on porous volcanic soils, the striped pyxie toad is found, whose only nurseries may be ephemeral puddles which quickly evaporate. It has therefore acquired several remarkable adaptations. Firstly its eggs develop very rapidly, hatching in two or three days. Secondly, the tadpoles, instead of hanging for several more days developing mouths, are ready to start feeding the next day. If conditions are good, the tadpoles may double their bulk in the first day of feeding, and nearly double it again in the second day. After that growth is slightly less rapid, but the metamorphosed tadpoles are ready to leave their rapidly shrinking puddle in about three weeks.

It often happens, however, that feeding conditions are poor, since there may not be much organic matter in temporary puddles. In this case the tadpoles band together in shoals and move slowly over the bottom, flailing with their tails to set up a current which will bring up all available food from the mud. At this stage many tadpoles die of overcrowding or are cannibalized, a biological necessity which ensures that others will develop faster. The banding together of the tadpoles also has the effect of deepening the part of the puddle in which they are congregated; what little water is left flows down into the depression, lessening the surface

Finally, the layer of tadpoles bakes into a crisp pemmican biscuit in the hot sun. With the next rain it will be converted into rich food for the next crop of striped pyxie tadpoles.

area and cutting down evaporation. Complete drying out of the puddle may be delayed sufficiently by this method for the tadpoles to finish their development. They also congregate in natural depressions, in the spoor of buffalo or antelope or in footprints of wading birds. But in really poor conditions the tadpoles are unable to complete their cycle in time. Thousands of them are stranded and the hot sun bakes them into a crisp sheet. But their lives have not been wasted. Their bodies become part of the material of the dry puddle bed, and the puddle that forms there in the next season's rains will be extremely rich in the most nutritious food for tadpoles—the bodies of dead tadpoles. Next year the tadpoles in this puddle will develop more rapidly on this diet and metamorphose at half the size so that their prospects of winning the race against evaporation are brighter than are those of tadpoles in puddles that did not dry up the year before.

Reptiles breeding

Rainfall governs the onset of breeding in the amphibia, for without rain those away from permanent water would have no pond or puddle nurseries. Rainfall is also the decisive factor governing the breeding of reptiles. In temperate climates where cold and short days inhibit breeding, warmth and longer days stimulate it. But in tropical regions, especially on the equator itself, where day length is constant and it is normally warm enough for breeding all the year round, the coming of the rains is the seasonal event that triggers it off.

Many kinds of reptiles are egg-layers—agamas, geckos, skinks, monitors, crocodiles, tortoises and terrapins. The ubiquitous side-necked helmeted water tortoise or terrapin is as mobile on land as in the water, and turns up in the merest rain puddle in the middle of dry bush, to feed on the tadpoles and insects. When danger threatens, it buries itself in the mud of the puddle bottom, sometimes heaving up and stranding tadpoles as it does so. All terrapins are carnivorous, but all tortoises are vegetarians. When the rainy season comes tortoises awake from aestivation to feast on seedlings and fresh green growth. But it is amazing that tortoises can still exist in much of their range: it appears barren even in the season of greatest growth, because the superabundant goats have stripped to bare twigs and earth all edible vegetation below goat-reach.

Tortoises and terrapins dig nests for their eggs in

Even such giant tortoises as this big adult African spurred tortoise disappear during the dry season, presumably hidden among boulders where their boulder-like carapaces escape notice. With the flowering of the thornbushes and the coming of the rains, they emerge from aestivation to feed and breed.

A baby Nile crocodile surfaces among floating water fern.

120

The male three-horned chameleon has a prehistoric look, but with his grasping feet, prehensile tail, leaf-like shape and colour, and rocking, wind-blown-leaf gait he is wonderfully well adapted for living in bushes.

A newly-hatched rhombic egg-eating snake among the cast wings of termites. Egg-eaters are non-venomous but in certain areas closely mimic the colour pattern of the highly poisonous night adder. In semi-desert areas where the night adder does not occur, the egg-eater appears to mimic other venomous snakes. It is not certain whether this is true mimicry, or convergent evolution: both species may have arrived at a similar patterning and coloration as the best camouflage for the area in which they live.

rain-softened ground. They also urinate copiously while digging, wetting the ground further. For them it is the actual softening of the ground by rain that is crucial in the onset of breeding, as is also the case with the Nile crocodile. After the start of the first rains, crocodiles mate and the female digs her nest in a shady rain-softened, sloping shore. The incubation period coincides with the following short dry season, during which the mother guards her nest against monitor lizards and other predators. Soon after the onset of the long rains the eggs begin to hatch. The baby crocodiles are very vocal, like baby birds, and their underground chirpings prompt the female to release them by digging up the nest. She goes on guarding her fifty or so hatchlings for at least six weeks. When she basks in the shallows at the edge of the lake the young climb out of the water onto her back and head. Periodically she submerges, leaving the babies swimming. When she resurfaces they climb back onto her. But in spite of her care there is very heavy mortality among the young crocodiles from monitors, marabous, eagles and other crocodiles.

Like chelonians and crocodiles the majority of lizards are egg-layers. Geckos cement their eggs in small clutches under bark or stones, skinks lay them under rocks, agamas may dig nest-holes in the earth. Most chameleons, too, are egg-layers and dig nests in the ground: the flap-necked chameleon excavates a hole with her hind feet as deep as she can reach while holding onto the rim. She lays thirty or forty eggs, and afterwards tamps the earth down and even scatters dry twigs and grasses over the nest to hide it. Some chameleons avoid the perils of descent onto the ground by giving birth to their young in the bushes. In East Africa chameleons found up mountains are viviparous. The two-lined or high-casqued chameleon is found at altitudes at least up to 3,000 metres on Mount Elgon, the Aberdares and other mountains, clinging to lichen-covered St. John's wort bushes, or among everlasting flowers. Lower down, at about 2,000 metres Jackson's chameleon is very common in places. The three-horned males look like miniature *Triceratops* dinosaurs, and in territorial disputes lock horns in a slow-motion trial of strength; the females are hornless. The advantage of retaining the developing young within the body in reptiles that live at high altitudes is that the female, by basking in the morning sun, or sheltering from too much heat at midday, can select the most comfortable temperature for herself, which is probably also the optimum one for the eggs.

The embryos are protected from desiccation or too much damp, and from bacteria and fungi, in a superbly camouflaged mobile incubator. The babies are born encased in a jelly membrane which sticks to twigs or leaves. They quickly struggle free of this wrapping, and set off with eyes swivelling and tiny hands clutching, perfect miniature replicas of the mother.

Among snakes, as among lizards, the majority are egg-layers. Some even show maternal care: the python and blind burrowing snakes incubate their eggs, remaining coiled around them for some weeks until they hatch. Some baby snakes may be mainly insect-eaters until they have reached a large enough size to tackle the normal prey of their species. The egg-eating snake, however, need not eat for several months after hatching and its first meal will probably be a gecko's or other small lizard's egg. As it grows it will climb trees to take the eggs of weavers and finches. Baby egg-eating snakes are, therefore, most likely to hatch before the breeding season for other reptiles and birds.

Nesting birds

Rainfall is the dominant factor in stimulating breeding in birds. For some, the sight of falling rain is enough, but insectivorous species wait until new foliage has brought an abundance of insects. Weavers, which need long grass for weaving their nests, wait until it has grown, but swallows, which use mud, can start building soon after rain. Where there are two rainy seasons in the year some small birds may breed in each.

The onset of the rains, therefore, is the signal for a great burst of breeding activity. The arid bush country is refreshed and springlike with new greenery and flowers, and every thorntree or thicket provides a nest site for at least one pair of birds.

The bush is full of bird noises, some harsh, some melodious and sweet. Among the most characteristic sounds are the songs of the duettists. Duetting is restricted to various tropical groups, and is unknown in temperate regions where female birds usually do not sing. In duetting species, both cock and hen sing. Their notes are different, not sung at random but alternated antiphonally in such a precise way that the calls sound as if they could only have come from one bird. Most duettists live in dense vegetation, and the calls help to keep the pair in contact.

The best known duettists are the bou-bou or bell shrikes. Some have a simple duet of only two calls.

The red-backed scrub robin delays its nesting and therefore the hatching of its chicks until insects are plentiful, a little later on in the rains. Both parents bring caterpillars, flies and small grasshoppers for their nestlings.

Day-old red-backed scrub robin nestlings are naked and helpless, but have remarkably strong necks to hold up their big heads. In contrast to the nestlings of most other birds, which gape in response to vibration, they gape vertically as soon as they hear the musical call of a parent bringing food.

The cock kori or giant bustard displays by puffing out the feathers of neck and under-tail, a conspicuous visual signal as he tours his territory on the grassy plain in the early morning.

D'Arnaud's barbets calling together. Duetting strengthens the pair bond; during the breeding season arid bush country rings with the noisy, jubilant calls of barbets.

The black-headed gonolek, with black back and brilliant red breast, skulks in thick bushes, the male calling a short whistling "Yoik" which is answered by a tearing-cloth sound from the female. In contrast the black-and-white tropical bou-bou has an extensive repertoire of antiphonal melodies, some tonally very pure and pleasing but varying from district to district. The cock may give a series of bell-like calls, the female answers with a harsher note. A pair rehearse their duets and together evolve new patterns which they do not share with other pairs. This means that if a cock calls and another bird responds correctly, she is his mate.

Some other bird groups that duet in pairs do so as conspicuously as possible, perched close together on top of a bush or a dead tree. Red and yellow barbets shout "Tweedle-de-tweedle" over and over again, while d'Arnaud's barbets sing a loud four note song perched side by side, one with tail cocked up, the other beak up, tail down. Pairs of red-billed hornbills call a monotonous and continuous "Wot, wot, wot". As the tempo increases, so their wings half open until at the end of the duet they look like burlesques of heraldic eagles. Fish eagles also call conspicuously, flinging their heads up and back, a flashing white signal which augments the sound. Their loud ringing call of three descending notes is possibly the most evocative bird sound of Africa.

Conspicuous and evocative bird calls of other areas are the loud croakings of touracos in the forests; the various mournful or plaintive cuckoo calls which are said to herald rain; the distinctive laughing or lamenting of doves in early morning; the bubbling of the coucal; the cackling of francolin or spurfowl in the bush; and the loud "Er-anna, er-anna" of the white-bellied bustard of the plains. In these the songs are more conspicuous than the birds' displays, but in others it is the displays that catch the attention. The greater flamingo gives a magnificent, majestic, wings-out display; on a diminutive scale it is echoed by the grey-headed kingfisher which simultaneously trills penetratingly. The kori bustard tours his territory with a stately walk, his tail cocked up to display the conspicuous white of fluffed under-tail coverts. Crowned cranes have an excited but stately dance which they perform while melodiously calling with a plaintive yet haunting sound. Lesser flamingos perform a parade in which a hundred birds march closely-packed with heads up and necks flushing bright pink, long red legs flashing, voices cackling and honking in communal excitement.

In some birds with particularly eye-catching displays the males have elaborate ceremonial plumages for the nuptial season. The bishops are glossy black with much brilliant red or yellow. Whydahs are not so colourful but have superb long tails. A bishop male displays over his territory, flying up with a peculiar undulating, floating flight, rump feathers puffed and giving aggressive sizzling calls. He may have half a dozen sparrowy females nesting in his territory. Jackson's whydah makes circular rings in the grass from which he displays. Some species spring into the air and fall back into the grass repeatedly, giving an extraordinary impression of ragged black objects being regularly tossed up. No whydah nest has ever been found; all species are thought to be brood parasites, some in the nests of waxbills. Female whydahs watch their hosts' nesting preparations and the sight causes them to come into breeding condition. They lay two or three eggs in the waxbills' nest, but unlike baby cuckoos, baby whydahs do not evict the host's own eggs or nestlings. The colour of the whydah's eggs, the mouth markings of the nestlings, and the juvenile plumages, mimic those of the specific host, and fledgling whydahs stay with their foster-family for some time after

The greater flamingos' nuptial display is a slow-motion ballet in which each bird walks with neck full length, turning its head smartly first to one side, then the other. Suddenly several birds together snap their wings open in salute. Right: Egg-laying is remarkably simultaneous in the nesting colony so, except where rising water has caused the odd egg to be abandoned, the chicks all hatch within a few days.

A crowned plover settles itself to brood newly-hatched chicks until their damp down has dried. While the chicks are small, the parents find food for them and call them to it, indicating it by pointing with the beak. Even when full grown and independent, the chicks may stay with the parent.

In some years the entire crop of baby greater flamingos is eaten by marabous within a week or so of hatching. The adults bravely attempt to defend their chicks, but are no match for the storks.

The male red-headed weaver uses pliant green twigs to weave his nest; other species may use grass blades. This bird has almost completed the foundation ring on which the globular nest will next be built up.

leaving the nest. The cocks pick up their foster-father's vocabulary and later use it in their own courtship.

Other birds with special nuptial plumes include three nightjars which have splendid elongations of wing or tail feathers. The long-tailed nightjar has elongated central tail feathers; the pennant-winged has very long primaries, almost twice as long as itself; but most extraordinary is the standard-winged nightjar which has in each wing a very long-shafted feather with a broad flag at the end, and in flight looks like one large bird followed by two tiny ones. In the breeding season the wattled starling moults its head feathers and grows extraordinary black wattles on crown and throat while the rest of the skin of the head becomes bright yellow. Some sunbird males moult into brilliant metallic plumages for the breeding season.

Nesting weavers are very conspicuous, the cocks brightly coloured in the breeding season with yellow or red breast off-set by black bib, cap or mask. The male builds elaborate hanging nests, beautifully woven of vegetable fibres, often grasses. He starts by constructing a ring at the tip of a branch, and from this hangs upside down to advertise for a mate by wing-flapping and calling in urgent buzz or chirping chatter. The nest is built up around the ring, a tough sphere with a side entrance, protected by porch or pendant tunnel. Weavers are polygamous, and may build nest after nest. Their colonies are a seethe of excitement as

the males flutter upside down, displaying yellow or red breasts or underwings to the sparrowy females. Other varieties of weavers build huge communal nests that look like hay stacks.

Greater flamingos' colonies are also the scene of intense excitement as nesting gets under way on some rocky island. Often at this time black clouds build up for the afternoon downpour, and winds whip up banks of spume along the shores. The flamingos on one island all lay almost simultaneously, and the young hatch within a week of each other. But within days of the first chick's appearance, marabous come planing down onto the island and begin gulping down the hatching chicks, in spite of brave protests from the parents. As one marabou becomes gorged, another takes its place, and between them the whole island may be cleared of flamingos in a very short time. Only a few young may be shepherded by their parents to islets in deeper water, and so escape.

Migrants

In much of East Africa the long rains occur during the Palearctic winter, so that the tropical season of abundance coincides with the temperate season of scarcity. Many European birds therefore migrate to Africa in the autumn, about 600 million arriving by way of the Mediterranean and trans-Sahara route. Harriers and eagles are funnelled down the Rift Valley, where they remain to feed

Six hundred million migrant birds arrive in Africa each Palaearctic autumn, many of them waders which descend in great flocks onto the Rift Valley lakes. Here a flight of ruffs alights among the reeds, where a female Bohor reedbuck is feeding and a flock of African spoonbills resting.

Thousands of European shovelers winter on the East African lakes every year. Drakes are often in partial eclipse plumage when they arrive, but most have moulted into their distinctive breeding colours before they depart in the spring.

mainly on rodents. Large eagles are grounded by rain. Just before a storm they perch on a small tree or fence post, and sit hunched and sodden-looking until after the rain has passed. But swifts and hobbies follow the storms; when black clouds build up they appear by the hundred, hawking flying insects brought out by rain. European swallows appear in huge flocks, and are very conspicuous when they first arrive, clustered in thorntrees, particularly near the soda lakes where there are abundant lakeflies for them to feed on. Vast flocks of waders also descend on the Rift Valley lakes. Sometimes migrant ducks, particularly hordes of shovelers from Europe, African pochard, Hottentot teal and Cape wigeon, congregate in rafts in uncountable numbers.

A local migrant that lives in a world of perpetual abundance is the glossy-black Abdim's stork. It breeds in the northern savannahs during the rainy season there, then moves southwards, appearing in East African grasslands with the November rains. During the next dry season it continues south with the rain belt then returns north with it and reappears in East Africa with the April rains. Thus it spends its whole life in the grasslands when these are wet and abundantly swarming with insects.

At one time Abdim's storks were able to feast on the swarms of locusts that also followed the rains. The migration pattern of locusts is particularly suited to the climatic regime of East Africa. They fly downwind in swarms towards frontal systems of converging air flows and so accumulate in the rainy areas. There, the flush of new vegetation provides good feed for the next generation of hoppers, who march along in bands, devouring every green thing. The migrating swarms used to do catastrophic damage to crops and all green vegetation, and were so numerous that although storks and other big insectivorous birds gorged themselves on them, they had no effect on the insects' destructive powers. Since 1944, however, the plagues have been kept under control by anti-locust organizations which prevent them building up in the outbreak areas. The multi-millions are now rarely seen in East Africa, only solitary individuals.

Other insects dependent on seasonal vegetation also follow the rains. The African migrant butterfly sometimes migrates on a locust-sized scale. Millions of white males flutter past like a snowstorm, a few sulphurous females among them. A small, nocturnal, brown moth, *Spodoptera exempta*, whose larvae are known as army-worms, migrate, flying with the wind, unaffected by the coolness of highland nights or the falling of rain. The females lay their eggs among freshly sprouting grass. When the caterpillars hatch, they eat their way across grassland in a living green-black carpet. Birds descend on the feast: storks, ravens, hornbills, as well as smaller insect-eaters, devour vast numbers of caterpillars, but there is such a glut of them that millions more remain. They graze good pasture to stubble and in more arid places do not even leave the stalks. Suddenly they all disappear underground to pupate. A further rainfall stimulates the emergence of the next crop of moths: forty-five million can take flight in a single night.

Insects

Another spectacular sight of the rainy season is the swarming of the termites. For many weeks winged sexuals have been waiting underground for the right conditions for their maiden and only flight. The rise in humidity with rain brings the right conditions. Over a wide area, all the nests of a given species swarm simultaneously, so that many millions of insects emerge together. The direction of flight is quite random and unless there is a strong wind to carry them the insects do not fly far. There is no attempt to mate in the air, so the flight is dispersal rather than nuptial. When the termites land, their wings easily break off at their bases. Now wingless, they run about searching for a mate, and then proceed to tandem, the male closely following the female, stroking her abdomen with his antennae. The pair search for a suitable cranny in which they can excavate the beginnings of a new termitarium. Only when this is done do they copulate.

Flying termites are supplied with fat and proteins so they can start new colonies without having to leave the shelter of the nest to forage. They are therefore very valuable as food for other animals and a great many creatures gorge themselves on them whenever they can. During a daytime emergence great vultures and tawny eagles toddle incongruously after them on the ground, with hornbills, storks, baboons and monkeys; while chanting goshawks, grasshopper buzzards, pygmy falcons, rollers, bee-eaters, shrikes, barbets, glossy starlings and others hawk them from nearby trees. At night Verreaux's eagle owls, pearl-spotted owlets, nightjars, and bats take the places of the diurnal animals. When larger predators are satisfied the lesser ones feast: geckos, toads, crabs, scorpions, rats, hedgehogs, shrews, and bushbabies by night; dwarf mongooses, skinks, elephant shrews, spiders, frogs, and ants by day. As termites swarm irregularly no animal can be entirely dependent on them, except perhaps the hobby, which like Abdim's stork moves south with the rain and then returns with it, feeding largely on flying termites on the way.

Ants also swarm at the start of the rains. Theirs is a nuptial flight, for they copulate on the wing. But the male takes no further part in the foundation of a new colony as he dies shortly afterwards. When she lands the female bites or shoulders off her wings and searches for a place to start a nest. Not all species of ants produce mating swarms; queen safari ants for instance, are wingless, though the males are those large brown wasp-like but harmless creatures known as sausage flies that bumble and zoom about lighted rooms at night during the rains. It is not known how the males find the queens. Sausage flies and other flying ants are pounced on during the glut by every insect-eater, habitual or opportunist.

Enormous population explosions of other insects occur during the rains. There are insects everywhere, in nearly every environment and exploiting almost every kind of food, though as with mammals, green vegetation forms their major food source.

Army-worm caterpillars rival locusts in their destructive powers in grassland. During an army-worm invasion, all insect-eating predators, such as this small skink, gorge themselves on the glut but have little effect in stemming the tide of munching caterpillars.

A royal pair of termites sets off in tandem after a very brief flight. The female has shed both pairs of wings, but the male still has a pair attached. They cannot have flown far before pairing, for the attendant workers and two castes of soldiers are still milling about.

Several pairs of royal termites may excavate a nest together in which they will found a new colony. When safely underground they will mate and the female will begin egg-laying.

Almost all butterfly and moth caterpillars, all grasshoppers and most crickets, many flies and beetles, are leaf-eaters. All bugs except a few blood-suckers suck plant juices. So with rain-stimulated plant growth there occurs a great proliferation of insects, whose ecological importance is enormous since they convert a good percentage of all plants into animal matter.

Insects are not only conspicuous through force of numbers, some are conspicuous through the sheer size of the individual. Nearly all exceptionally big insects are tropical because only in a warm climate can long respiratory systems work efficiently. Some very large insects occur in East Africa: swallowtail butterflies in the forests; huge saturnid moths in the bush; and very big scarabs or dung beetles.

Dung beetles emerge from their underground cocoons at the beginning of the rains, and soon set to work burying dung. They can be seen homing onto fresh dung almost as soon as it hits the ground. They fashion the dung into balls much larger and heavier than themselves, and trundle them along by running backwards rotating the dung-ball with their hind legs. When a suitable spot of soft ground is reached, each beetle buries itself and its dung-ball in an underground chamber. There the beetle shreds the dung, remoulds it into a beautiful pear-shape or oval, and lays an egg in a hole in the top of it. The dung remains moist underground, and the larva, when it hatches, eats away the inside of the pear and then pupates inside the remaining thin outer shell.

Some dung beetles are compulsive dung-buriers and bury all they can find. Individuals of a small species less than 2 centimetres long can bury 100 cubic centimetres of dung each in a night. The work of dung beetles is therefore of great importance in the ecology of a region. Underground the dung is acted on by bacteria and transformed into substances useful to plants; left above ground it dries and remains useless. By burying dung containing seeds (page 96), the beetles assist their germination. In addition, their work is of prime sanitary importance, since many diseases are transmitted by faecal matter and faeces-visiting flies.

The very biggest dung beetles, which are stout, well-armoured, almost cubic creatures about 7 centimetres long, would seem to be large and tough enough to be left alone, but they are eaten by sacred ibises which dig them up from their underground burial-chambers and swallow them whole, and by lilac-breasted rollers which beat them to a pulp on an anvil-stone first. Augur buzzards and pale chanting goshawks strip off the wings and elytra and swallow only the abdomen, leaving the head and thorax to tramp steadily on as if nothing had happened. Smaller dung beetles are the staple diet of banded mongooses, who follow in the wake of feeding herds of elephants and buffaloes and rake over their dung to find the beetles. There is even a plant, *Hydnora*, that "eats" small dung beetles. A parasite on the roots of acacias, its presence is unseen until huge red beetle-trap flowers burst open at the surface of the ground. Dozens of beetles are attracted by the flowers' foetid scent, crawl on the smooth waxy petals and tumble down inside. They are unable to climb out of the slippery trap and quickly perish, to be digested as the flower wilts and disappears below ground again.

In the forests, butterflies may be on the wing at any time: gorgeous *Charaxes*, orange, iridescent blue, mother of pearl, bask on sunlit foliage in the glades or speed among the trees with powerful, direct flight. But at this season even arid bush comes alive with butterflies: big yellow swallow-tails, orange-brown monarchs, small sulphurs that cluster in brilliant patches on damp mud, magenta-tipped whites and orange-tipped lemons, tiny long-tailed blues, peacock-eyed pansies. In the hot sun, even the smaller ones fly much faster than the butterflies of temperate regions. When the sun goes behind a cloud, they touch down and hide among foliage to wait for the shadow to pass. In cooler cloudy weather butterflies at a salt-lick close up when the sun goes in; when it comes out they open wide and bask.

One of the most interesting among the many remarkable butterflies of East Africa is the gaudy commodore, which parallels the great contrast between the wet and the dry seasons by having distinct seasonal forms. In the dry season the butterfly is reddish-orange and brown to match the predominance of these colours in the dry season landscape. In the rains it is predominantly blue, chequered with black and red. Intermediate forms also occur, purple-blue with a broad red band, and both wet and dry season forms may be seen on the wing at the beginning of the seasons.

There are so many plant-eating insects that, if they are not completely to dominate the area, they must be controlled. Spiders are a check to some extent on insects. They are an immensely varied order, second only to the insects themselves, and because they are all carnivorous their effect on other invertebrates is profound. Very many kinds of spider use their silk to weave elaborate webs in which to snare insects many times their own size. Others, such as baboon and wolf spiders, live in burrows in the ground from which they emerge at night to leap on their prey, as the tiny jumping spiders do by day. Yet others, such as crab spiders, lie in wait, often beautifully camouflaged in a flower, hidden from their own enemies and from their prey.

But probably the greatest natural check on the hordes of plant-eating insects are the parasitic wasps. The females tirelessly search for individuals of the particular insects which are hosts to their species. When a female wasp finds a host she pierces it with her ovipositor and lays one or many eggs inside its body. When the larva or larvae hatch they feed within the host, first on non-vital tissues,

When conditions are right for a termite emergence, they are often right for flying ants also. With ants it is a true nuptial flight, as mating takes place on the wing.

All sorts of insect-eating animals feast on the nutritious flying termites. European storks, winter visitors to Africa, are among the birds that take termites whenever there is an emergence.

Dung beetles are of great ecological importance, for they are compulsive buriers of dung. Females trundle balls of it backwards until they find a suitable spot of soft ground in which to bury it.

An African monarch or milkweed butterfly ostentatiously spreads itself before its maiden flight, displaying bright colours that warn of its poisonous nature. The female of the diadem, an unrelated and innocuous butterfly, mimics the monarch and so is protected from predation, but the diadem male is an unprotective velvety-black with a round, white, violet-edged patch on each wing.

so that the host appears to develop normally. Later they attack the vital organs, then emerge from the empty skin, spin a cocoon and pupate.

Many parasitic wasps are extremely small, among the smallest known insects. Their larvae develop within a single egg of the host, or within a tiny host such as an aphid. Some species are counterchecks on other invertebrate predators, and may be very large. The spider-wasps are black or metallic blue insects up to 3 in (75 mm) long, with long legs and iridescent blue wings. Their prey is almost exclusively spiders; the very big species prey on baboon spiders, which are stung so accurately in the main nerve centre they remain alive, though paralyzed, for forty days, providing fresh food for the wasps' larvae. Large spider-wasps bury baboon spiders in the ground; smaller spider-wasps are potters, making vases of cemented mud which they stock with stung spiders for their larvae to feed on. Other potter wasps provision their beautiful little pots with paralyzed caterpillars. Some parasitic wasps reproduce by parthenogenesis; in others a single egg develops into a multitude of individuals. The phenomenon of parasitism is immensely complicated, and is made even more so by the fact that many of the parasitic wasps are themselves the specific hosts to other parasitic wasps. For example the velvet ant (actually a wasp whose flightless, black and white females run about on the ground and can give a very painful sting) parasitizes another species of wasp (striped blue, black and yellow) which lives in scattered colonies on sandy, sloping shores of soda lakes and stocks its burrows with a supply of blue-bottle flies.

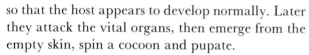

A close-up of the inside of a *Hydnora* beetle-trap flower. Small dung beetles, attracted by the smell, have tumbled down inside and cannot climb out again up the waxy, overhanging petals.

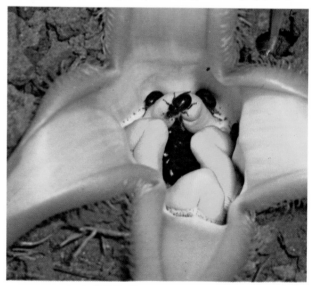

Other stinging, biting and parasitic invertebrates also proliferate during the rains, to the discomfiture of many animals, from tortoises to lions and elephants. Mites and ticks abound. After rain, hungry ticks—larvae, nymphs or adults—can be seen on the heads of tall grasses waiting to climb onto any large animal that brushes past. Ticks can survive very long periods away from their hosts, even in arid places. Some show almost incredible powers of water conservation, and can

The tiny yellow jumping spider is diurnal, and stalks its prey by stealth, catching it by leaping onto it. Relatively huge prey such as bluebottle flies, are quickly subdued by injections of paralyzing venom from the spider's fangs.

The crab spider can assume the exact coloration of the background on which it is sitting, here a cluster of milkweed flowers. It sits motionless with front legs held wide, waiting for an unsuspecting insect, here a small blue butterfly, to alight.

Red, yellow or orange combined with a dark colour advertise that an insect is poisonous, and birds quickly learn to ignore one flaunting these colours. The lubber grasshopper, unlike most of its family, is slow, clumsy and conspicuous. It is extremely unpalatable, exuding a caustic yellow fluid from joints in legs and thorax when disturbed. It enhances its warning by flashing its wings open, displaying large areas of colour and abdominal stripes. Among bees and wasps, bold stripes are common.

The big wolf spider lives in a burrow lined with silk, and emerges at night to run actively after its prey. The female carries her eggs in a cocoon attached to her spinnerets. When the baby spiders hatch they swarm all over her body and she has to keep them off her face and headlamp-like eyes by frequent windscreen-wiper movements of her pedipalps.

live for ten years or more without food or drink. Whilst on their host they gorge themselves on its blood, then drop off to the ground again. One female may lay 4 to 8 thousand eggs, which probably implies the chances of finding a host are slim. Once attached to the host, ticks are very difficult to remove, and cannot easily be dislodged by scratching or nibbling. Primates, with their nimble fingers, are uniquely able to rid themselves and each other of such parasites. Large animals rely on tickbirds.

Tickbirds or oxpeckers are the sole exploiters of a unique ecological niche. There are two species, the yellow-billed and the slightly smaller red-billed. The ranges of the two overlap, and they are not clearly separated ecologically; mixed flocks can sometimes be seen feeding together on the same animals. They visit a wide variety of game from rhinoceroses, buffaloes and giraffes down through zebras and antelopes to warthogs, and man's domestic stock. Some pastoral people welcome

Buffaloes resting at midday tolerate the red-billed oxpeckers that work over their hides, removing engorged ticks and wiping the blood off their beaks on the buffaloes' great horns.

European swallows cluster on lakeside trees before setting out on their return journey.

Antelopes are much troubled by the swarms of biting, blood-sucking flies—*Stomoxys*, tsetses, horse flies—during the rains. This kongoni bull has been deliberately kneeling down to rub his horns and face in a mud puddle—an action which probably has sexual significance. After scratching his flanks with his horns, he twists round to gnaw an itch on his rump.

them, but today oxpeckers have disappeared from many areas where modern farming techniques involve the regular use of cattle dips. When ticks disappear from cattle, and game is no longer abundant, the tickbirds also disappear. They obtain the whole of their sustenance from the hides of their hosts, mostly by picking off engorged ticks with their specially-adapted flattened beaks. It is not so much the ticks themselves which provide the birds' food, as the blood which they contain. The diet is supplemented by blood-sucking *Stomoxys* flies, and by the blood and tissue taken direct from wounds on the living animal. These are consequently often prevented from healing for a long time, and even enlarged. Surprisingly, ox-

peckers are tolerated remarkably by their hosts, even when they peck at raw wounds. Sometimes such wounds, especially those on the shoulders or flanks of rhinoceroses, are infected with microscopic parasitic worms (*filaria*). The oxpeckers may be disinfecting the wounds to some extent rather than feeding on healthy tissue; the worms prevent healing rather than the oxpeckers. Often their attentions are actively solicited, as when an antelope stands stock still with its ear presented for deticking. The tolerant mammals derive further benefits from the symbiotic relationship since the tickbirds warn them of danger; and in return the birds, as well as finding food, warmth and a mobile display ground, are provided with readily-obtainable hair with which to line their nests.

The end of the rains

At the end of the rainy season the showers become fewer and lighter, separated by longer rainless periods. Antelopes may be up to the withers in a sea of tall grasses above whose seeding heads the growing calves are hardly visible. At this time greater kudus give birth, and hide their new calves in the dense vegetation. Many plants have seeded, and bushes and trees are now about to lose their leaves to conserve moisture in the coming dry season. Some take on brilliant reds and yellows before leaf-fall, like trees in temperate regions in autumn. Small birds may still be feeding their fledglings, possibly the young of a second brood. Those birds that nest in the dry season are making their scrapes or collecting nesting material now. Others, which have finished breeding, go into post-nuptial moult: weavers, bishops and whydahs into a sparrowy plumage; nightjars, wattled starlings and some sunbirds into plumage almost indistinguishable from that of their females. The European migrants gather in flocks before they depart. Lakeside trees are loaded with swallows, European storks spiral in the thermals in flocks of five hundred or more before setting off for their breeding grounds. Butterflies feed on late flowers, grasshoppers munch drying grassblades. In the higher rainfall areas, all is lushness and late summer abundance, but in arid country the grass has already turned to standing hay, bushes and shrubs have dropped their leaves and wear the bare winter look typical of the dry season. Small animals are hidden underground, rivers are low and water holes drying: the country begins its next long season of drought.

The tiny lilies that flowered at the start of the rains now bear big, succulent fruits. The leaves always appear cropped, probably by antelopes, when they first push out of the ground and after the flowers have wilted. The stem collapses under the weight of the ripening fruits.

The African lion, the superpredator and lord of the plains, calmly surveys his domain. Yet he could face starvation if he is not able to hunt and feed without constant surveillance from tourists.

Epilogue

Almost nothing can be written about wildlife today without sounding a note of doom, for the dangers that confront ecological systems and the wild animals in them are many and real. Pollution of the environment is still only an emerging problem in East Africa, but the greatest threat is the rapid growth of its human population and the very limited nature of its resources. East Africa has great open stretches of land, but because of its extreme climate most of this is not good arable land, while primitive farming methods have already exhausted much of the soil. Land pressure is growing steadily as towns expand and farms spread. In the most fertile areas big ranches that produced food for the many are broken up into subsistence plots for resettlement by the few. In less fertile areas, people, cattle and maize spread at the expense of wild animals, trees and grass. The wild animals have been squeezed into East Africa's sixteen national parks and into a score of smaller game reserves. The national parks cover an area of about 32,000 square kilometres, but this is only three per cent of the land area of Kenya, Tanzania and Uganda combined. And even vital parts of the national parks and forest reserves are liable to be handed over to the people for settlement. As human populations build up around and even within the parks, the menace of poaching increases, free movement of animals on great traditional migration routes is impeded, and possibly the entire ecology endangered.

However, the revenue from wildlife and the interest shown in it by visiting tourists have convinced many governments that the animals and wildernesses must be preserved, and that a living antelope or zebra is worth more to the country as a tourist attraction than its carcass is worth as meat. The half-million tourists who visit East Africa's parks each year stimulate local economies and bring in much-needed foreign currency. But conservationists are now worried that the flood of visitors is having a harmful effect on the animals, particularly in some of the most popular parks. The great cats, especially, are subjected to nearly continuous observation by tourists, and may spend much of their daylight hours in the centre of a ring of motor vehicles. Cars may chase alongside them when they are trying to hunt, and they may even be prevented from making their kill or from eating it. The cumulative effect of visitor impact may be that lions, cheetahs and leopards are unable to get enough to eat. Similarly, game viewing too close to waterholes, particularly in the dry season, may prevent game from getting enough to drink. In the breeding season, the eggs of ground-nesting birds are at high risk on mudflats and short-grass swards, particularly in the vicinity of the spectacular and much visited soda lakes. Even the vegetation suffers from visitor impact. It only needs two or three cars to follow in the track of another across a piece of grassland for the wheels to leave permanent scars on the pasture.

The management of wildlife in East Africa, as anywhere else in the world, begins with the management of people, both indigenous and visiting. The obligation of the African governments to produce economic prosperity for their peoples and more land for human settlements must be balanced by plans for land-use that are more scientifically determined, with more thought for the morrow. There is a growing awareness of the value of wildlife in East Africa so it is very much to be hoped that governments will adopt far-sighted plans already proposed to leave major parts of ecological systems undisturbed, while making other parts available for game ranching and viewing.

Bibliography

BERE, R: *The Way to the Mountains of the Moon*, Arthur Barker, London, 1966

BERE, R: *Antelopes*, Arthur Barker, London, 1970 and Arco Publishing Company, New York, 1970

BROWN, L: *Africa*, Hamish Hamilton, London, 1965

BROWN, L: *African Birds of Prey*, Collins, London, 1970

CALDER, N: *Restless Earth*, British Broadcasting Corporation, London, 1972

CLOUDSLEY-THOMSON, J L and CHADWICK, M J: *Life in Deserts*, G T Foulis, London, 1964

COATON, W G H: "Association of Termites and Fungi", *African Wild Life*, Vol. 15, No. 1 (March 1961), p. 39

COE, M J: "The Biology of *Tilapia grahami* Boulanger in Lake Magadi, Kenya" *Acta Tropica Seperatum*, Vol. 23, No. 2 (1966)

COLE, S: "Ancient Mammals of Africa", *New Scientist*, Vol. 394 (June 1964), p. 606

COWIE, M: *The African Lion*, Arthur Barker, London, 1966 and Golden Press, New York, 1966

DORST, J: *Before Nature Dies*, Collins, London, 1970

DORST, J and DANDELOT, P: *A Field Guide to the Larger Mammals of Africa*, Collins, London, 1970

GROVES, P: *Gorillas*, Arthur Barker, London, 1970 and Arco Publishing Company, New York, 1970

HALL, R: *Discovery of Africa*, Hamlyn, London, 1970

JARVIS, J U M and SALE, J B: "Burrowing and burrow patterns of East African mole-rats *Tachyoryctes*, *Heliophobius* and *Heterocephalus*", *Journal of Zoology*, Vol. 163, Part 4 (April 1971), p. 451

KYLE, R: "Will the Antelope recapture Africa?", *New Scientist*, Vol. 53 (March 1972), p. 640

NAPIER, J R: "Profile of Early Man at Olduvai", *New Scientist*, Vol. 386 (April 1964), p. 86

NEAL, E: *Uganda Quest*, Collins, London, 1971

WAGER, V A: *The Frogs of South Africa*, Purnell, South Africa, 1965

WILLIAMS, J G: *A Field Guide to the National Parks of East Africa*, Collins, London, 1967

WILLIAMS, J G: *A Field Guide to the Birds of East and Central Africa*, Collins, London, 1963

WILLIAMS, J G: *A Field Guide to the Butterflies of Africa*, Collins, London, 1970

Glossary

Aestivation Dormancy during heat and drought.

Agaric Umbrella-shaped fruiting body of a fungus.

Antelope A ruminant characterized by cylindrical annulated horns; a name loosely applied to all gazelles and related forms.

Biomass Total weight of living animals to a given area.

Boundary effect The concentrating of animals where two vegetational zones meet.

Brood parasite Animal (particularly bird) that lays its eggs in the nest of another species which then rears the adopted offspring.

Browser An animal, especially an ungulate, which feeds on the twigs and leaves of shrubs and trees.

Bush Loose term for any East African countryside that is not farmland or forest; dry country with fairly dense bushes and small trees, but usually with some grass grading to wooded and open savannah in higher rainfall areas, or to semi-desert (sub-desert steppes) in lower rainfall areas; also known as Nyika ("wilderness").

Characins Very diversified group of fishes recognized by small adipose fin on the back between dorsal fin and tail; occur in Africa and tropical America.

Chelonian Group name for tortoises, terrapins or turtles.

Conidia Small fruiting bodies of fungi produced by constriction of specialized structures, conidiophores.

Convergent evolution The development of similar characters in animals and plants belonging to different groups.

Ecology The study of the mutual relationship between animals and their environment.

Feral Animal or plant that has escaped from domestication and reverted to living in the wild.

Gazelle A small to medium-sized, slender, graceful antelope, with long legs giving great speed (e.g. gerenuk, Grant's gazelle, impala).

Gymnosperm Plant whose seeds are not enclosed in a true ovary (e.g. conifers).

Hygroscopic Readily absorbing or retaining, or becoming coated with, moisture.

Individual distance The distance between individual birds of flocking species, maintained by the distance at which they can only just peck one another.

Insulberg Huge isolated granite boulder; "island rock".

Larva Immature stage of insects in which the young do not at all resemble the adults cf. **nymph** (see below).

Mimicry Resemblance in colour and/or structure as a means of camouflage or self-protection.

Nymph Immature stage of insects in which the young resemble the adults except for the absence of wings cf. **larva** (see above).

Ovipositor A specialized structure, usually a long tubular organ particularly of female insects, for laying eggs in a suitable place.

Palearctic A division of the Earth which comprises Europe, Asia north of the Himalayas, northern Arabia and Africa north of the Sahara.

Parthenogenesis Reproduction without fertilization by the male.

Plankton Floating or weakly swimming, usually very small, plants or animals in a body of water.

Protozoa Single-celled or non-cellular animal organism.

Pupa Inactive third or chrysalis stage of insects that have larvae; often enclosed in a cocoon.

Rift valley Elongate depression on Earth's surface produced by vertical displacement.

Ruminant Member of a sub-order of even-toed ungulates which chew the cud.

Savannah Tropical African grassland, open or dotted with small trees (tree savannah); corresponds to "steppe" in Asia and "prairie" in North America. In South Africa, temperate grassland is known as "veld".

Symbiosis Living together of organisms (two animals, two plants or plant and animal) in mutually beneficial partnership.

Thermal An ascending column of warm air.

Ungulate Hoofed mammal; now restricted to the Perissodactyla or odd-toed ungulates (e.g. zebras, rhinoceroses) and Artiodactyla or even-toed ungulates (e.g. hippopotamus, pigs, giraffes, antelopes).

Viviparous Giving birth to live young in contrast to laying eggs.

Index

Numbers in italics refer to illustrations

138

storm, behaviour during, 111-12, 126
stream, mountain, 23, 24
sunbirds, 32, 176
  bronzy, *Nectarina kilimensis*, 105
  scarlet chested, *Chalcomitra senegalensis*, 33
swallows, 40, 122
  European, *Hirundo rustica*, 127, 134
swamps, 45-9, 47, 48
swarming, 126-8, 129, 131
sweating, 84
swifts, 34-6, 127
  alpine, *Apus melba*, 32
symbiotic relationships (*see also* food chain), 54, 135

T

tadpoles, 32, 117, 118-20, 118, 119
temperature (*see also* climate):
  in dry season daytime, 80
  of flamingos' nests, 82
  of hot springs and lakes, 36, 38
  regulation in:
    birds, 32, 84, 84
    crocodiles, 82
    mammals, 84-6
termitaria, 56-7, 59, 59, 60, 60, 61, 67, 96, 107, 115
termites, *Macrotermes*, 56-61, 59, 115, 128-9 129, 131
Termitomyces (termite fungus), 107, 107
terrapin (water tortoise)
  helmeted, *Pelomedusa subrufa*, 120-1
thorntrees, *see* acacia
tickbird, *see* oxpecker
ticks, 132-5
tiger, water, 117, 117
tigerfish, *Hydrocynus*, 43
tilapia:
  mouth brooder, Mozambique, *Tilapia mossambica*, 44, 44
  T. grahami (Lake Magadi and Lake Nakuru), 36, 40, 113
  T. spilurus (Lake Naivasha) 45
toads, 80, 118-20
  African common, *Bufo regularis*, 118
  striped pyxie, *Pyxicephalus delalandii*, 119-20, 119
topi, *Damaliscus korrigum*, 60, 65, 66, 67
tortoise, 80, 94, 120-1

African spurred, *Testudo sulcata*, 120
hinged, *Kinixys belliana*, 69
water, *see* terrapin
tourism, 136
trade, early, 14-15
trees, *see* vegetation *and common names, e.g.* acacia
tribes, 26, 67
tsetse fly, *Glossina*, 69
turaco, Hartlaub's *Turaco hartlaubi*, 50
turtle, *see* terrapin, tortoise

U

Uaso Nyiro River, 112, 113

V

vegetation:
  damage by elephants to, 54, 96, 96-7, 107
  effect of fire on, 93-5, 98
  of:
    bush, 52-4
    high mountain slopes, 24, 28-9
    montane forest, 24, 24-8
    plain, 62
    riverine forest, 50
velvet ant, *Dasylabris deckeni*, 132
velvet mite, *Trombodium*, 113, 114
volcanoes, 17, 22-4, 25, 37
vulture, prehistoric form, 12-13
vultures, (*see also* lammergeier), 76
  hooded, *Necrosyrtes monachus*, 76, 76
  Rüppell's griffon, *Gyps ruppellii*, 36, 76, 77
  white-backed, *Pseudogyps africanus*, 76

W

wallowing, 86-7, 87
warthog, *Phacochoerus aethiopicus*, 42, 60, 82, 87, 90, 100
wasps, parasitic, *Melanopa* and other genera, 130-2
water:
  conservation, 52, 87-90, 100
  consumption, 86-8, 90
  content of vegetable matter, 88
  frequency of visits to, 86-90. 95, 98, 100
  loss, for cooling, 84
  transport by sandgrouse, 90,

waterbuck, *Kobus*, 42, 65, 70-1
  defassa, *K. defassa*, 42
waterhole, (*see also* puddles), 87, 92
  behaviour at, 86-7, 87
  maintenance of, 97, 115
waterlily, *Nymphaea caerulea*, 45, 48
water tiger, 117, 117
waxbill, *Estrilda* and related genera, 124-5
weaver, 47, 122, 126, 126
  red-headed, *Anoplectes melanotis*, 126
  social (communal nest builders), *Pseudonigrita*, 126
web of life, 75
wells, artificial, 101
Western Rift, 17, 25, 34-6, 44-5
wet seasons, (*see also* breeding, nesting, swarming), 6-7,
  animal behaviour in, 104-8, 111-12
  appearance of aestivators, 114-15
  end of, 135
  floods in, 111-13, 112
  flowering in, 104-7, 105, 106, 107
  germination in, 112, 112
  migration in, 108, 126-8
  rainfall in, 104
wheatear, Schalow's *Oenanthe lugubris*, 34
whydah, *Ploceidae*, 124-5
  Jackson's, *Drepanoplectes jacksoni*, 124
wildebeest, *Connochaetes taurinus*, 65, 67, 85, 87, 92, 95, 95, 106, 108
winds, 54, 92-3, 104
woodlice, 115

Z

zebra:
  Burchell's, *Equus burchelli*, 67, 92, 100 101
  Grevy's, *E. grevyi*, 72, 73
zebras:
  adaptation to drought, 100
  calving of, 108
  coloration of, 84
  dust bathing, 92, 91, 92
  galloping to rain, 106
  grazing of, 65
  habitat, 65
  migration, 95, 108
  prehistoric form 12-13
  resting of, 82

HIDDEN AGENDA

By the same author

THE BLACK UNICORN

THURSDAY'S CHILD

A TIME TO SPEAK

A CAGE OF HUMMING-BIRDS

WELCOME, PROUD LADY

CABLE CAR

THE SABOTEURS

THE GANTRY EPISODE

THE PEOPLE IN GLASS HOUSE

FAREWELL PARTY

BANG! BANG! YOU'RE DEAD

THE BOON COMPANIONS

SLOWLY THE POISON

FUNERAL URN

THE PATRIOTS

I SAW HIM DIE

SUCH A NICE FAMILY

THE TROJAN MULE

THE BLUESTOCKING

JUNTA

THE UNSUITABLE MISS PELHAM

BURDEN OF GUILT

THE IMPOSTOR

HIDDEN AGENDA

June Drummond

VICTOR GOLLANCZ

LONDON

First published in Great Britain 1993
by Victor Gollancz
A Cassell imprint
Villiers House, 41/47 Strand, London WC2N 5JE

A catalogue record for this book
is available from the British Library

ISBN 0 575 05629 0

Photoset by Rowland Phototypesetting Ltd
Bury St Edmunds, Suffolk
Printed in Great Britain by
St Edmundsbury Press Ltd, Bury St Edmunds, Suffolk

For the good companions of the bad years.

Glossary

bitter-einders:	people who fight to the last
CCB:	Civil Co-operation Bureau
donner:	hammer/kill
induna:	headman
Inkosi:	Lord. Term of respect
Magtig:	Almighty
Mlungu:	white man
muti:	medicine
ou boet:	old boot/old pal
ous:	old mates/pals
skelm:	rogue
twak:	rubbish/shit
umfaan:	young boy
verneukery:	trickery/double-dealing

1

'Trumpeters,' said Todd Scallan.

The hornbills had erupted from the wooded gorge about a kilometre north of the ridge where he sat, and were tumbling across the sky in fantastic gyrations, emitting wild braying cries.

Todd's companion, a young Zulu, looked up from the portable primus he'd set on a slab of stone and squinted at the birds.

'Like politicians,' he said. 'Squawk, squawk.'

Todd grinned, watching the hornbills swing eastward over the plain, a basin bisected by a tributary of the Klip. The river bed was almost dry, the grassland pallid with drought. Several white-owned farms dotted the rolling land, which showed here and there a patch of brilliant green where the farmers had found enough borehole water to irrigate their fodder fields. The largest farmhouse and the brightest green lay directly below the ridge, between the gorge and the river.

Behind him rose the hills, all Zulu territory once. A road had been cut in preparation for the building of the new dam. The foundation of the dam-wall already stretched across the V-shaped cleft by which the river left the hills; but the economic recession had stopped work two months ago. The ground on either side of the cleft lay scoured clean of vegetation and the earth showed red and raw.

Pieter du Toit had been up in arms about that. 'I told the buggers they can't put a dam there. That rock's frot, man, full a holes like cheese. Another storm like we had in eighty-seven, that whole lot's going out to sea.'

Todd saw in vision the floodwater coursing down the valley, tearing at its flesh, carrying down sand and boulders, trees and livestock, smashing through the farmlands, destroying everything in its path.

'Like Noah,' he said aloud.

Thulani Msomi smiled, but said nothing. He was growing used to Todd's elliptical remarks. Some people said Todd was crazy, the way he talked to himself or seemed not to hear what was said to him. Thulani knew it was because he thought deeply and stored everything in his head.

He was careful about what he said and what he did. You could tell by the way he worked with a plant, first making drawings or photographs of it where it grew, then taking specimens to store in his special boxes. He would write down the Latin name, then the common name, and then he would ask Thulani whether it was named by the Zulus, and what were its special uses. Thulani's great-uncle Siphiwe, who was a noted herbalist and prophet, said that Mr Scallan was a very clever man.

Thulani picked up a battered aluminium pot and climbed down the rocky slope to the nearest pool in the river bed. He filled the pot and carried it back, leaping easily from one foothold to the next. He had the athletic build of the Nguni race, slender but strongly muscled, the long torso encasing a powerful heart and lungs. He sang as he moved, crooning the same phrase over and over. At boarding school he sang the latest pop tunes, but here on his home ground he stuck to traditional songs. It was his way of living in two worlds.

Regaining the ridge he lit the stove, placed a can of beans

and one of vegetable curry in the pan of water, and set it on the primus to heat.

Todd, who had been leafing through a pack of photographs, handed them to Thulani and said, 'How do you call these?'

Thulani studied the pictures. Some showed plants strange to him, others he knew. He tapped a picture of a small tree starred with small mauve flowers and green four-lobed fruit.

'That grows on the edge of the forest,' he said. 'I don't know its name. My uncle soaks the bark in hot water, and puts it on wounds.'

Todd retrieved the photograph and made a note on its back. Thulani continued to examine the pack. After a while he said, 'This grows here.'

Todd craned sideways to look. 'That's a cycad,' he said, 'but I don't expect you'll ever see that particular species . . . not in the wilds.'

'I have seen this plant,' insisted Thulani. 'With spikes, like a porcupine.'

'You've seen its first cousin, *Encephalartos natalensis*. That's quite common in these parts, but the species in the picture is *Encephalartos woodii*, and the only known specimens are in cultivation. I took that photo in the Botanical Gardens in Durban. It's a male plant. There are no known females.'

Thulani stared. 'Then this will die?'

'Not necessarily. Cycads grow suckers that become new plants. Like the sugar-cane. But out here, yes, it's finished.' Todd had been speaking Zulu but he switched to English. 'Extinct in nature,' he said, 'like the dinosaurs.'

The water in the pan was boiling, and he lifted out the tins of food, opened them, shared the contents on to two tin plates and handed one to Thulani. They ate in silence, wiped the plates clean with bread and replaced them in Thulani's pack. Todd preferred to carry his specimen-tins and camera himself. Thulani took care of his sketching block, the primus and the food.

They shouldered their rucksacks and set off along the path that led past the head of the gorge. Below them the plain shimmered in the three o'clock heat.

'Will it storm?' Todd asked. Thulani shook his head.

'The hand is still tight,' he said. It was a vivid description of the weather that had gripped the country month after month.

They passed the point where the hornbills had broken from the forest, but saw no birds stirring. The stream that would normally tumble into the gorge was bone dry, its bed cracked like crocodile hide.

Rounding the shoulder of the hill, they came in sight of the flatcrown tree where Todd had parked his station-wagon. A second vehicle was drawn up in the thin shade; a truck with a metal canopy. Two men stood next to it. One was tall and swag-bellied, dressed in khaki shirt and trousers, with a soiled felt hat on his head. The other was dark and slightly built, dapper in jeans, Hawaiian jacket and ankle boots. Both men carried rifles.

Todd slowed and raised a hand in greeting. The fat man took a step towards him, lifting the muzzle of his gun.

'Mister, this is private property. You trespassing.'

Todd smiled pleasantly. 'I'm sorry. I have a permit from the magistrate in Dundee, allowing me to work on the dam site.' He found the slip of paper and held it out. 'I'm doing an ecological survey of the plants that will be submerged when the dam fills. There are valuable grasses in these valleys. If they extend over this sector, we can—'

The fat man cut him short, scowling. 'I know about the grasses, they my grasses. This is my land and I say who comes on it, not the magistrate. The road to the dam's through the hills.'

'I know, I usually use it, but today—'

'May I see?' The man in jeans spoke suddenly, reaching out a hand for the permit. He read it and gave it back to Todd.

'Seems OK, but you should be more careful, Mr Scallan. People are jumpy about trespassers these days.' His dark eyes searched Todd's face. 'You probably don't know that the Smedleys were murdered last night.'

Todd shook his head in shock. He remembered Mr Smedley: an old man, rubicund and loudmouthed, who'd farmed the land west of the village. He'd been in the store lunchtime yesterday, talking to Piet du Toit.

'I hadn't heard. We've been up here since dawn.'

The dark man nodded. 'A gang shot their dogs, gunned down Smedley at his barn, broke into the house. They made Mrs Smedley unlock the safe and gun cupboard. When she tried to get to the intercom, they cut her throat.'

'Do the police have any leads?'

'They're looking for a couple of labourers Smedley fired last month.'

The fat man snorted, staring at Todd with bright porcine eyes. 'Any kaffirs come near my place at night, I shoot first, ask questions after.' He looked at Thulani. 'This your boy, mister?'

'My assistant.' Todd's voice was cold, and the dark man said quickly, 'It's on the permit, Fanie. Mr Scallan and assistant.' He gave Todd a half-apologetic smile. 'Where are you making for now?'

'The hotel.'

'Take the lower road, then. Keep away from the tops. Whoever killed the Smedleys has their guns and ammo, and the police are after them. You don't want to get caught in the crossfire.'

The fat man was already walking back to his truck. Todd

13

signed to Thulani to get into the car, and turned to the dark man.

'Thanks for the warning. Perhaps I'll see you later in the pub?'

The other shook his head. 'I don't live here. I'm an agent for fertilizer, just passing through. You follow the lower track, it'll bring you to the hard surface about two kays along. That takes you straight to the village.'

As Todd made to climb into the station-wagon, the dark man called after him. 'Bit late, aren't they, with their survey? I'd have thought they'd have done it before they started digging.'

'There were delays. Government wheels turn slowly.'

'Yeah.' The man raised a hand in casual salute and walked away to the truck, his rifle cradled in the crook of his arm.

Thulani watched him go. 'Why does an agent for fertilizer carry a gun?'

'Scared, I suppose.' Todd scanned the hills above them. 'Maybe we should be, too.'

2

During the school holidays, Thulani lived with his grand-parents on a smallholding north of Muller's Drift. Todd met him each morning on the parking lot behind the hotel, and left him there at the end of the day's work; but this after-noon, with the Smedleys' murder fresh in his mind, he took the boy right to the foot of the path that zigzagged up to the Msomis' land, and watched him until he reached the hedge that marked the boundary of the kraal.

The hotel was a sprawling bungalow that dated back to the days when there'd been a glint of gold in the drift. On Friday nights most of the younger farmers stopped off at the pub for a few beers. Todd dumped his pack in his bedroom and went to join them.

The main bar was crowded. The drinkers clustered together, their faces sharp with anger and unease.

'Why the Smedleys?' the barman said. 'Everyone knows the old man was on the bones of his arse, sold up weeks ago, didn't have a thing worth stealing.'

'They took cash and guns,' a farmer said.

'Ach, man, twenty-five rand and an old army rifle! If they wanted real money, why didn't they hit Pienaar's place, or McFarlane's, or just about any other farm? Bert Smedley told me last week he even sold his furniture. He was going to take his missus up to Jo'burg, to live with their daughter.' He filled a tankard and spun it along the counter to a customer. 'An' that stuff about them being killed by their own blacks, that's twak. The only two left at Westbrook was Smedley's induna and his son, and they not the sort to murder anyone.'

'How can you tell who's the sort?' demanded the stationmaster. 'These days, you can't trust any blacks. They kill each other all the time, why would they mind killing us? I'm glad I live in the village. I wouldn't like to be like you ous, out in the sticks.'

The argument rose. Todd looked round for Piet du Toit, the owner of the store, and saw him making his way to the veranda that overlooked the street. Todd followed and found the big man leaning his elbows on the rail, head bent. He glanced up as Todd approached.

'Man,' he said, shaking his head, 'these are terrible times.'

'Did you know the Smedleys well?'

'A long time. They were customers and friends. Old Bert lent me some cash once, to tide me over a bad patch. Lately I been carrying him. He was a good farmer, but the drought finished him.' Du Toit stared across the road to the railway track where a few trucks waited for tomorrow's load. A wan light shone on the empty platforms and the scrubby gumtrees behind the ticket office. 'I'll miss the old bugger,' he said.

'Have there been other attacks in the area?'

'Ja.' Piet straightened up and turned to face Todd. 'Five months ago, Tiny van Vloten was shot. That time it was his labour, everyone knew. Tiny treated them like dogs, and one day they fixed him. But the Smedleys were good employers, paid well, ran a school for the farm kids. Now, they gone. I tell you, the whole country's run mad. Killing, wherever you look. I never used to carry a gun. I do now – everywhere I go.'

'It's a time of great change. Change breeds unrest.'

'Everyone blames politics; the other man's politics. The

ANC blames Inkatha, Inkatha blames the ANC, everyone blames the government. There's always politics, Todd. That doesn't mean you can go out and donner people who don't think the same way as you. Magtig . . . I don't know . . .'

Todd remembered his encounter on the hill. 'I met a farmer this afternoon, up on Crown Ridge. Big fellow, big paunch on him, squinny-eyed. He told me he owns the land above the gorge.'

'That'd be Fanie Kramer. He leases the high land, uses it for summer pasture. His main place is on the plain, along the river.'

'There was another man with him; small, dark, fancy dresser? He said he was an agent for fertilizer.'

Piet shook his head. 'Agents come and go. They mostly deal direct with the farmers, not through me.' He gave Todd a sharp look. 'What were you doing on Crown Ridge? It's way off your beat.'

Todd said vaguely, 'I was checking the grasses . . . and there were hornbills over the gorge.'

Like Thulani, Piet was accustomed to Todd's cryptic turn of phrase. He accepted the explanation without comment, and, soon after, strolled away to his house behind the general store.

Todd was about to return to the bar when he heard his name called softly. He leaned to peer into the shadows below the veranda rail and saw a figure there, biblical, wrapped in a blanket and leaning on a tall staff.

Todd said in Zulu, 'Msomi? Is that you?'

The figure moved forward. Light shone on a wrinkled face, eyes filmed with cataract.

'I greet you, Todd.' The old man's voice was deep and resonant. 'If you please, I wish to speak about Thulani.'

'Of course. Come up.'

'It would be better to talk in the yard.'

17

Todd fetched two tankards of ale and walked with his guest to the yard behind the kitchens, where there was a bench. Hosea Msomi was seventy-three years old, dark and spare as a wild pear tree. He had put on a suit for this formal visit and wore a knitted cap pulled down over his ears. The blanket about his shoulders smelled of woodsmoke and tobacco.

He settled himself on the bench and for a time drank and exchanged the proper courtesies. After a while he drew a folded paper from an inner pocket and handed it to Todd.

'I brought you Thulani's school report,' he said. 'There is also a letter from his headmaster.'

Todd read the report first. It spoke in glowing terms of Thulani's scholarship, his sporting ability, his leadership qualities.

'He's done very well,' Todd said. 'We can be proud of him.'

The old man signed to Todd to read the letter. Its message was direct:

Dear Mr Msomi,

I am writing to you as the head of Thulani's family. As you see from the enclosed report, he has distinguished himself in all fields of our school life. If over the next few years he maintains this excellence, he will be eligible to apply for a bursary to attend university.

That does not mean he will succeed. Thousands apply each year for a very few bursaries. If he does succeed, it will mean that he must study for years without remuneration.

Before I speak to him, or put his name forward, I must be sure that I have the approval of Thulani's family. I feel I should tell you that on the last occasion that Thulani's mother visited us, we talked about university for him and she told me there are difficulties in the way. I gained the impression that her husband, not being Thulani's father,

*is against giving the boy more than he can give his own
children. He has said that when Thulani leaves school, he
should find work and contribute to their education.*

*I told Mrs Gwala that I understood Mr Gwala's feelings,
but that if Thulani secures his degree he will be in a far better
position to help than if he has only a matriculation certificate.*

*My deep conviction is that Thulani is a gifted child and
should be given every chance to develop those gifts, not only
for his own and his family's sake, but for the country's. God
knows we will need all our skills if we are to come through
the testing years that lie ahead.*

*You, as the family head, will make the final decision.
Perhaps you will think over what I've written and let me
have the benefit of your advice. If you care to discuss things
with me, please let me know, and I will make the necessary
arrangements. I can drive to Muller's Drift, or send a car to
bring you to the college.*

*It remains only for me to repeat that it has been a pleasure
and a privilege to have Thulani at Lauder. Whatever
decision is reached, I am sure his future is bright. It is in
his nature to strive, and to succeed.*

I look forward to hearing from you,

> *Yours sincerely,*
> *Harold Storton,*
> *Headmaster, Lauder College.*

Todd folded the letter and handed it back.

'My wife and I,' said Hosea, 'have read this many times.
We feel honoured by what Mr Storton has written, but there
are things to be considered.'

'There are, indeed, father.'

Hosea spread gnarled fingers. 'First,' he said, 'there is the
matter of my daughter's husband, Gwala. You know that he

is not Thulani's father . . . He ran off to Igoli. Gwala married my daughter but does not accept Thulani. Thulani belongs to me and my wife now. Gwala pays nothing for him, although he's a rich man. He has taxis in town, he has a good house in Umlazi, he drives a big car, but he will give nothing to Thulani.'

Todd said quickly, 'I'll speak to my father. If Thulani gets a good matric, the Trust will help him.'

'Thank you; but the help must not all come from outside. Some must come from me and my family. I have money saved. My oldest son Philip has done well, he's with a bank in Johannesburg, and his own children are earning now. I've written to him about Thulani. He wants to meet the boy. Will you do me a favour? The next time you go to Jo'burg, will you take Thulani with you?'

'Of course. Where does your son live?'

'In Soweto.' Hosea fumbled in the folds of the blanket once more. 'His full name is Sibongile Philip Msomi. This is his address, and the telephone numbers at work and at his home. Thulani can stay with him while you are in Igoli and come back here when you do.'

'Fine. Have you spoken to him about all this?'

'Not about the bursary. I won't raise his hopes until all is settled.' Hosea sighed. 'When he was a small child, he spoke of becoming a doctor. Now it's a doctor of plants, like you. That would please me. Our family has a tradition of healing through herbs, and such work would keep him closer to us. Doctors of people go to the towns where there are big hospitals and big money.'

The old man broke off and sat with head raised, listening. 'Are there cars on the hill?' he said.

Todd scanned the slopes north of the village. The moon was rising, giving the peaks a spectral sheen. Over to the west, car lights moved along the contour road.

'Yes,' he said. 'Two cars.'

'Police,' Hosea said. 'They're looking for Cele. They came today to question us about him. They think he killed Mr and Mrs Smedley. I told them, it was not Cele who did that killing. It was not one of our people. It was others from outside, from far away.'

Todd didn't ask why Hosea believed that. He had no intention of being drawn into a local murder hunt.

'I'll be going to Jo'burg next weekend,' he said. 'I'll take Thulani and drop him off at his uncle's house in Soweto.'

Climbing the path behind the hotel, Hosea arrived at the outlying huts of his family settlement. They were built in the old tradition: beehive-shaped, with neat thatch held in place by plaited withies.

This was the kraal of his younger brother. A dim light glowed in the main living-hut, and a pungent smell from within it suggested that Siphiwe was boiling up one of his herbal brews. He did a good local trade in medicines, and also sent roots, bark and animal parts to the store owned by his cousin Jacob in Maritzburg.

There was a flourishing market in such remedies. Even Hosea's eldest daughter Miriam, who was a sister at King Edward VIII Hospital in Durban, admitted that when Western treatment failed, the doctors were not above calling in a herbalist or sangoma to talk to the patient.

Siphiwe was a stickler for the old ways. He still stored his grain in the clay-lined pit in his cattle byre. He still furnished his hut with hand-carved wooden pillows, plates and spoons. He observed all the ancient taboos. No one who had been contaminated by recent contact with death might drink sour milk in Siphiwe's house, or scatter the seed-corn in his fields.

This attitude raised problems with Hosea's wife, Ellen, who was town-bred and a Christian, as well as being extremely strong-minded. Five years ago, at her insistence, a man had come from town to show the people how to lay a carbon bed to purify the spring water, and build a tank to store it in. Siphiwe had been dead against the scheme.

The water, he said, belonged to the ancestral spirits, and if it was tampered with the crops would fail. Hosea, while he certainly didn't wish to offend the ancestors – he always poured a little of his beer on the ground for them, and often left scraps of food for them to nibble during the night – decreed that the project must go forward.

Now Siphiwe insisted that the angry spirits had inflicted the drought as punishment; while Ellen claimed that without the tank all the families on the hill would die from lack of water. The tank was kept padlocked to prevent theft by the shack-dwellers in the valley, and two of Hosea's dogs were tethered near it every night. They rose to greet him now, whining as he bent to touch their heads. They were fatter than most kraal dogs, for Ellen was fond of animals and bought them pellet-food from the store.

Hosea walked on past his cattle byre, and turned through the gate in the fence that surrounded his house. This was square, solidly built of concrete blocks with a corrugated iron roof. Ellen had supervised the building, giving the workers many words when they failed to meet her exacting standards. It was a fine house, warm in winter and cool in summer, with a vegetable garden where Ellen grew maize and cabbages, beans, sweet potatoes and pumpkins, and the small-leafed spinach Hosea loved.

Now that their children were launched in the world, Thulani had his own room when he came home from school. When he was away, Ellen kept her knitting-machine there. She made jerseys, caps and shawls which she sold to that same cousin Jacob's store.

Every cent she could save went into the building society

for Thulani. He was the apple of her eye, this grandson who could hold his own with the white boys from rich families, and who was going to make a great name for himself one day.

Tonight she met her husband at the gate, greeted him and said at once, 'Did you talk to Mr Scallan? Will he take Thulani to town?'

Hosea nodded, at the same time holding up a calming hand. 'Yes, yes, it's arranged for next weekend, but we mustn't talk of it too much. We mustn't give the boy false hopes. Let him visit Philip and pay his respects. If he makes a good impression . . .'

'He will! He's a fine boy.'

'Philip's a businessman,' said Hosea drily. 'He won't put down money unless he can be sure of a good return.'

'Thulani will pay him back, with interest.'

'You believe that and so do I. We must make Philip believe it. I will write to him tonight, and during the week you must make a sack of vegetables for him. People who live in town like vegetables.'

They began to walk towards the house. As they climbed the steps to the narrow porch, Ellen shook her head. 'A university is a dangerous place. If he goes there, Thulani may become involved in politics, he may get into a lot of trouble.'

Hosea shrugged the blanket from his shoulders and gave it to her to fold.

'He may be safer at university,' he said, 'than he is here, with us.'

3

The man coming down from the mountainside moved at a steady pace, his climber's boots making a clinking sound on the drought-hardened footpath. Behind him towered the escarpment that marked the frontier of Lesotho, and below, crouched in a fold of the foothills, was the straggling village where he planned to spend the night.

He wore faded army trousers, a windproof jacket over a checked woollen shirt and a brown cloth cap with ear-flaps. He carried a backpack and a rifle fitted with telescopic sights.

His face was somewhat broad at the cheekbones, the fea-tures regular. His skin glowed copper red in the slanting light, despite its protective coating of barrier cream. The hair that curled at the nape of his neck was bleached almost white. He was not tall; but a certain tension in his body, the near-squinting concentration in his Siamese-cat eyes, warned that he might be a dangerous man to cross.

He reached the settlement, skirted a group of wattle-and-daub houses, passed the trading store and entered the square blockhouse of the post office. Its two staff members stopped talking at the sight of him. One returned to his post at the switchboard. The other remained leaning on the inquiry counter. He made no attempt to greet the newcomer, but thrust a slip of paper towards him.

'Caller phoned three times,' he said. 'Says it's urgent.'

The fair man studied the number written on the paper, then thrust it into an inner pocket.

'Did you recover the stock?' demanded the postmaster.

The other shook his head. 'It was over the border twelve hours ago.' He turned away.

'You can use our phone,' offered the postmaster.

The fair man ignored him, continuing on his way to the door. The postmaster seemed about to call him back, but changed his mind.

'Let the bugger pay for his own calls,' he muttered.

'Who wanted him?'

'Who cares?'

'He's a good tracker.'

The postmaster made no answer. Licking a forefinger, he began to count a pile of banknotes. The stiffness of his back indicated that he did not propose to discuss the matter further.

In the ranger's hut he rented, the fair man went to the wall telephone and dialled the number he'd been given. A voice said at once, 'Oswald.'

'Luther. You called.'

'Yes, I want to see you.' The voice was phlegmy, punctuated by deep rasping breaths. 'Come at once. Tonight.'

'Get stuffed. I've got a job to do.'

'That's been arranged. I'll expect you by six tomorrow morning.'

Colour flooded the fair man's face. He said flatly, 'I don't take orders from you. You don't have authority.'

'I don't need authority. I have information. I'll use it if you try to buck me. Your job won't last long!'

The line went dead before the fair man could reply.

He sat for some minutes staring ahead of him. Then he dragged a suitcase from under the bed, threw in a few possessions and snapped the locks. He picked up his backpack, rifle and the case, and walked through the kitchen to the shed where he kept his car.

On Friday, four days after Luther received his summons, the upper concourse of Jan Smuts Airport outside Johannesburg was invaded by Muslims.

Every available space was occupied by white-clad pilgrims bound for Mecca. The men wore white skull-caps, the women white veils that wrapped their heads closely, hiding their hair but not their faces. For every pilgrim there was an attendant group of well-wishers. Children swirled round the outskirts, tumbled over piles of hand luggage, giggled and shrieked in a fever of excitement.

Round the main information desk people pressed five deep. Arrangements had been upset by trouble in the Gulf, and the uniformed receptionists battled to deal with a stream of questions.

The Tannoy intoned the departure warning, and like a flock of geese the pilgrims took off, some sweeping towards the security checkpoints, others hovering for last-minute farewells.

At the desk, the confusion dwindled. Passengers for other destinations, who had held back from the crush, now approached. It was not until a lull occurred some half an hour later, that one of the receptionists noticed the envelope. It was plain white, without stamps. The name MR NEIL SCAL-LAN was typed on it and below that, the words, TO BE COL-LECTED.

The receptionist shrugged and put it with the other messages in the rack behind her.

'I'm wasting my bloody time! I don't belong in this circus!'

Neil Scallan strode along the corridors of the Kempton Park Trade Centre at a furious pace. His hunched shoulders, jutting nose, full lips and prominent eyes conveyed an impression of primordial energy; a god about to take the form of a bull.

His companion, who was having difficulty keeping up with him, said sharply, 'Nonsense, you're invaluable. If COVISA can't come up with ways to end the violence, then all the rest is a waste of time.'

'You don't need me, man! This place is lousy with lawyers, not to mention politicians, security experts, military experts, academics, dozens of eggheads from the UN and the Commonwealth.' Scallan cast a disparaging glance at the main conference hall which they were now passing. 'It's a talking shop, Gordon. I loathe talk, I loathe decisions reached by consensus, above all I loathe constitutional law. I have a pile of work waiting for me. I want out.'

Gordon Leveret caught hold of Scallan's arm and forced him to halt.

'You listen to me. You're not here to argue constitutions, you're here to help us draft a new amnesty law. In a few weeks' time this country goes to a crucial election. One of the key problems – perhaps the key problem – is how a government of national unity will deal with the past. How will we indemnify the criminals of the past, the terrorists, the people in the police or the armed services who've committed crimes against their political enemies? If we get it wrong – if we draft a bad law – there'll be no hope of reaching a political truce. The country could slide into anarchy. You can't deny it, Neil.'

'Hell, there are dozens of better people . . .'

'No. There aren't. They may have your legal skill but they don't have your political credibility. The ANC trusts you because you've defended their activists. The National Party respects your experience and your international contacts. You are indispensable to us.'

27

As Scallan was about to reply, a messenger barred his way.

'Telephone for you, Mr Scallan.'

'Take a message, will you please?'

'The caller says it's very urgent, sir.'

'What's his name?'

'A Mr Wivern, sir.'

Scallan stared, then made a jerky movement of the head. 'Very well. I'll take it in the communications room.'

He moved along the corridor, Leveret scurrying at his side. They entered an office containing tables and chairs, closed- and open-circuit television, typewriters, word-processors and a fax machine. One of a bank of telephones rang and Scallan lifted the receiver.

'Scallan.' He beckoned to Leveret, who came close and bent his head to catch what he could of the conversation. There was a crackling on the line, then a voice said rapidly, 'This is George Wivern, Mr Scallan. You remember me?'

'I'm not likely to forget the name.'

'Listen carefully, I don't have much time. I'm on my way out. I've left something for you at the information desk at Jan Smuts Airport.'

Scallan's face was devoid of colour. 'Is this about David?'

'I can't talk. The desk at Jan Smuts. Make it fast.'

The line went dead. Scallan waited a moment, then replaced the receiver. He looked at Leveret.

'Did you hear?'

'More or less. It's some kind of hoax, Neil.'

'It could be Wivern.'

'Did you recognize the voice?'

Neil hesitated. 'It was a bad line. There was an aircraft taking off in the background.'

'That could be faked.' Leveret spoke with quiet urgency. 'Wivern is dead. His death was presumed by a court of law years ago. This is nothing but a cruel hoax.'

'Only one way to find out.' Galvanized, Scallan was heading out of the room towards the main entrance doors. 'Make my apologies to the chairman, will you?'

'I'm coming with you. You don't know what's behind this. You're a potential target.' Leveret broke into a trot to keep up with his companion. The prospect of travelling to Jan Smuts in Scallan's Porsche alarmed him, but he knew he had no choice in the matter.

The receptionist at the information desk produced the envelope at once and handed it to Neil.

'Who left it?' he asked. She gave him a rueful smile.

'I'm sorry, I don't know. There was a crowd leaving for Mecca; they were all round me. When they'd gone I saw the envelope on the counter.'

Scallan opened it. It contained a slip of paper on which was typed:

Colonel Hermanus Silbermann	R24,000
Sergeant Lemuel Moroka	R 7,500
Mr Mark Alberton	R17,000

Scallan stared at the list for so long that Leveret shifted uneasily. Scallan turned the paper for him to read.

'Silbermann and Moroka were with Wivern on the police team that investigated David's death,' he said. 'Alberton was a student at Wits, in David's time. He posed as a radical.'

'Posed?'

'Yes. After David's death it was established Alberton had been working as a police informer on campus. He was expelled from the university. Soon after, he and his family emigrated to Australia.'

'You think he sold David out?'

'Someone did.' Neil's initial shock was turning to rage. The sight of those names brought the taste of bile to his mouth. He had hated them all: the hulking, brutish Silbermann; Moroka, the black man who had betrayed his own kind; Wivern, fair-haired, specious, always with a smile on his face. He'd hated Wivern most of all. Aloud, he said, 'A hit squad.'

'What?'

'Those figures suggest a hit squad. Payment for work done.'

Leveret shook his head confusedly. 'If Wivern was on that kind of payroll, why would he tell you?'

'God knows.' Neil was trying to recall the exact words of the telephone conversation. 'He said he was on his way out. I took that to mean out of the country.' He swung back to the desk and beckoned the receptionist.

'Miss, the flight you mentioned, to Mecca? Has it left?'

'Yes, sir, over an hour ago.'

'Does it stop at Harare, Entebbe, anywhere in Africa?'

'Khartoum, only.'

'Is there any way of finding out who was on the passenger list?'

'We don't reveal that kind of information, I'm afraid.'

'But someone in the airport has it?'

The girl nodded. Scallan's urgency had communicated itself to her, and she said uncertainly, 'You could ask at the air controller's office, but I don't think they'll be able to help.'

Neil thanked her and rejoined Leveret. They began to move across the concourse, but before they'd gone far, Neil changed his mind.

'Wild-goose chase,' he muttered. 'Wivern needn't have been on that flight. In the past hour he could have left for Singapore, the UK, any damn place. Needn't have left at all.'

'What are you going to do?' Leveret sounded as nervous as he felt.

'Talk to Major Joubert.' Scallan tucked the envelope and its contents into his breast pocket and started for the exit.

Hurrying beside him, Leveret said, 'You must be careful, Neil, not to compromise our work. While I understand your concern about David . . .'

'The motive for David's murder was political,' Scallan said shortly. 'That brings it well within our terms of reference.'

Leveret sighed. Meeting Scallan's gargoyle stare, he thought he discerned Neil's true reason for wanting to pull out of the amnesty committee. In the matter of his son's death, would he ever be able to forget, or forgive?

4

'Is it wise,' said Anna Delport, 'to open old wounds?'

She had raised herself against the pillows to look down at Neil, who lay with a forearm across his eyes, still half-asleep.

'Is it?' she repeated, and he stirred.

'I've no choice.'

'And Meg? How will she take it?'

He made no answer. His wife had been in the United States for nearly two months, staying with their eldest son. She would return refreshed, bright, resolute in her opposition to Neil's wishes.

Anna slid out of bed and walked naked across the room to the bathroom. She had a fine body, slender and well-proportioned, and she was amazingly unconscious of it. She had no artifice, sexual or otherwise. Though she could be as passionate in her love-making as in her opinions, there was always a certain detachment in her, a need to consider and evaluate.

He'd learned that the first time he met her, at Bob Lanergan's place. It was one of those university gatherings, neither a meeting nor a party, the room filled with pot and rhetoric. He'd noticed Anna standing by herself in a corner, her blonde head tilted a little, her expression quizzical. He caught her eye and smiled, and to his surprise she came straight across to join him, threading her way through the arguing groups.

Holding out her hand, she said, 'I've been wanting to talk to you. I'm Anna Delport.'

'Neil Scallan.'

'I know. I've seen you in court.' She set down her empty glass, waving away his offer to refill it. 'No, I'd rather talk. I've been sitting in on the Magabe trial. Isaac was a student with me. Will you get him off?'

He was about to give her the stock answer, that he couldn't discuss a client with her; but meeting her steady gaze he found himself saying, 'I'm afraid it's unlikely. The State has a very strong case. We'll plead mitigating circumstances, of course.'

She said impatiently, 'I hate that phrase. A corrupt system isn't an extenuating circumstance for a crime. It is the crime. Isaac's whole life has been deformed by what the State is pleased to call Law and Order. His accusers should be in the dock, not him.' She gave a half-shrug and said, 'I know you're doing your best for him, more than anyone here.'

He glanced round at the earnest debaters in the room. 'It counts for something to speak out . . . to define principles and teach them to others.'

'It's not enough.'

'What is enough? What do you believe is enough?'

She thought for a moment. 'To oppose injustice by every means short of violence.'

In the years that followed she was twice detained by the security police, once for a few days and once for four months. Neil was by that time established as the leading defence counsel in political trials, and his intervention helped to secure her release; but her marriage broke apart, Delport not being a man to buck the authorities. Her parents, wealthy and conventional, stood by her but found her a sad puzzle. After her divorce she came to Neil and told him she was going abroad for a while.

'I'm off to see the outside world,' she said, smiling. The phrase, so often used by South Africans, always amused her.

'You'll come back?' He was surprised to find how much he wanted that.

'I hope so,' she said.

For three years he heard nothing of her save vague rumours that she was in England, America, Germany. Then one spring morning she strolled into his office in Jeppe Street. He could remember exactly what she wore – a cream skirt, a thick caramel sweater, suede boots. Her skin was rosy and her eyes sparkled.

'I'm looking for a job,' she said. 'Can you oblige?'

She worked for him until she was given a junior lectureship at the University of the Witwatersrand. That same year they became lovers, though she resisted it for some months.

'You're married with three children. Do you plan to divorce your wife?'

He tried to explain about his marriage. He'd met Meg soon after he was called to the Bar. She was the daughter of a senior advocate, surrounded by the wealth and status that position provided. Not that that was his reason for marrying her. He'd fallen in love with her delicate prettiness and her butterfly changes of mood. It had taken him a year or two to understand that she was deeply neurotic, prone to feverish highs and bouts of depression. When she was ill she clung to him desperately, threatening to kill herself if he left her.

'She's eating you alive,' Anna said. 'I don't intend to be eaten with you.'

Neil took her hand and held it against his cheek. 'I know I'm offering you a one-sided bargain.'

She'd accepted on her own terms. She would take no financial help from him. She bought a flat in Killarney, an easy distance from the centre of the city, and he visited her there from time to time. Their friends accepted the situation. Inevitably Meg learned of it, and was content to turn a blind eye to it as long as her own territory was not invaded.

Her life was centred on her children. Nothing, no one, had any meaning for her except in relation to the boys. She saw Neil as a good provider and a reliable father, but she was not interested in his work, his beliefs, or his feelings. She accepted that he'd made a success at advocacy, but she'd have been far happier to see him engaged in what she called 'ordinary law'. She was glad that Neil now found his physical pleasures elsewhere. Anna Delport was not getting anything Meg wanted.

Looking back, Neil saw that though Anna had been excluded from a large part of his life, she had exerted a profound influence on his children. His oldest son Alistair, who was devoted to his mother, showed his hatred of Anna by rejecting everything she stood for. He became a slavish supporter of the apartheid system, grew rich in the network of parastatal finance spun by the government, and was now an unofficial lobbyist in Washington DC.

David, the middle son, took after Neil. He chose to study law at Wits, and quickly became involved in radical politics. Unable to take his friends to Meg's home, he brought them to Anna's flat. He regarded her as an ally and a confidante.

The car bomb that killed Dave also blew his family apart. Alistair blamed Neil and Anna for encouraging his brother in what he called subversive activities. He refused to attend David's funeral, saying that he wasn't going to stand alongside a mess of fist-waving commies. Neil's response was to forbid Alistair the house.

Meg, torn by grief and tormented by the knowledge that David's killers would in all probability escape justice, retreated into a world of make-believe.

Every photograph of David, every letter, every reminder

of his existence, was banished from her sight. She would not allow his name to be mentioned. She had no comfort to offer Neil or twenty-year-old Todd.

One evening when Neil arrived at Anna's flat he found her sitting in the living room with a large photograph album on her lap. She handed it to him saying, 'Todd brought it this afternoon. He said he saved it for you.'

The book had a white cover, on which was written in flourishing script, 'The Life and Times of David Malcolm Scallan'. Inside were snapshots of David with the family, with friends, with his labrador Roly, with his school sports teams. Loose between the pages was a newspaper photograph of Neil on the steps of the Central Court with David and Wilson Ndhlovu. Looking at the picture, Neil had to rub tears from his eyes.

'This damn country,' he said. 'This damn world.' Holding the album to his chest, he muttered, 'Good old Todd.'

'I asked him in,' Anna said. 'I suppose it was wrong of me.'

'No, not at all.' Neil moved to pour himself a whisky. 'Meg's no use to him.' He turned to face Anna, glass in hand. 'Did you talk about Dave?'

'Yes, and other things. Todd told me he wants to be a botanist.'

Neil sat down to re-examine the photographs. 'Yes,' he said vaguely, 'he likes that sort of thing.'

'I promised to introduce him to Prof Malherbe at the Harleigh Institute.'

'Good. You do that.'

So, almost casually, Todd became Anna's protégé. She understood better than anyone his complex nature: his reticence, his stubborn convictions, the odd mix of reason and faith by which he lived.

It was Anna who introduced Todd to Aline Marceau, the

daughter of friends in France. Aline was not at all the sort of girl Meg approved of, being foreign, Catholic, an artist and without money. However, Todd married her and the marriage proved a happy one.

They had only one child, Joseph, who was born with a hole in the heart and needed a good deal of expensive medical treatment. Todd's job with the Harleigh Institute for Ecological Studies was prestigious but poorly paid, and Aline's pictures were not yet attracting big prices. Meg suggested that to save money they move into part of the Scallan house in Parktown. Anna urged Todd to accept.

'You're away a lot,' she said, 'and it's really not safe these days for a woman and child to live alone. Joe will have comfort and security. He can walk across the road to school.'

'What if Aline fights with Ma?'

'She won't. Todd, it's worth a try. If it doesn't work out, you can move.'

The plan was a success. Joe revelled in the fine old house and garden, and his health improved. Aline was able to give more time to her painting. Above all, Meg took on a new lease of life. Having a grandchild to fuss over and spoil brought her out of her isolation. For Joe's sake she took up the threads of her life again; ran the house efficiently, attended church, played her part in the social round.

Only those close to her understood how tenuous was her hold on normality.

Neil knew Anna was right when she warned against opening old wounds. To revive the inquiry into David's death could destroy Meg utterly.

Anna came back into the bedroom wrapped in a towel, her hair dark from the shower.

'Will Todd be home this weekend?' she asked.

Neil nodded. 'He'll leave Muller's Drift early tomorrow.

Aline expects him back at about nine. Meg's plane gets in at three, so we'll all go out to meet her.'

The prospect of reunion troubled him, and he rolled out of bed and went to take his turn in the bathroom. When he emerged, Anna had dressed and was in the kitchen brewing coffee. The portable television was switched on, but without the sound. On the screen, black figures ran silently across a dusty road, pausing to hurl stones at an unseen opponent. Flames belched from a row of shacks.

'Where is it?' Neil said.

'Richmond.' Anna leaned to turn off the set. Feeding bread into the toaster, she said, 'I don't believe Wivern's alive.' Then, illogically, 'Why would he get in touch with you? It's crazy. You should ignore the whole thing.'

He sat at the kitchen table. 'You really believe that?'

She stood tight-lipped, and he repeated, 'Anna? Do you believe I can ignore it?'

'No,' she said at last. 'I see you have to act. What will you do?'

'I've made some appointments.' He took the mug of coffee she offered him and cradled it in his hands. 'Did you ever know a student called Mark Alberton?'

'No. He was in the arts faculty, I think. Myra Rosenwert might be able to help you.'

'I'll ask her. Do you know if the university would have kept a file on David?'

'I suppose it's possible, in view of the circumstances.'

'Will you find out for me?'

'Yes. What exactly are you looking for?'

Her expression was almost hostile, and he looked away from her.

'I don't know,' he said, then added half to himself, 'I'm looking for peace and quiet, for God's sake.'

5

The house was a red-brick box, the fence surrounding it of moulded concrete, unpainted.

The fair man parked his car on the vacant lot next to the property and climbed out, looking about him.

This was the back-end of Benoni. To his left the narrow street quickly expired in uncompromising veld. To his right was a string of run-down buildings interspersed with yards crammed with scrap metal.

The fair man advanced to an iron gate and pressed the buzzer of the security control attached to the fence. A husky voice answered. He gave his name and the lock on the gate was released. He stepped through and advanced up an uneven path between patches of desiccated lawn, climbed a flight of steps and waited at the front door.

This was presently opened by a black woman in a pink overall. Her feet were thrust into worn slippers and her head was uncovered, showing tufts of grey hair round a shiny bald patch. She offered no greeting but hipped away towards the back of the house, leaving the fair man to shut the door.

He crossed a dark hallway and tapped on the door to his right, entering without waiting for a response. The room was large, probably designed as the main living room, but used now as bedroom-cum-office.

The bed was old-fashioned, with a carved oak headboard and a white cotton quilt. Alongside it was a cheap pine wardrobe and chest of drawers. On the bedside table was a telephone and the remote control of the security system.

The opposite side of the room was taken up by a rank of filing cabinets, a desk, a fax machine and a word-processor. A second telephone stood on the desk.

In the centre of the floor stood a trestle table covered with piles of newspapers arranged by title and date. Their pages were ruffled by the cold air flowing through two open sash-windows. Next to the left-hand window was a wheelchair in which sat a small, squat man with a tartan rug over his knees. He lifted a hand and waved the newcomer towards him, saying in a wheezing voice, 'Well, Mr Wivern, you did very well!'

'My name is Luther. Use it.'

'Sensitive, are we?' The squat man chuckled, the sound sliding into spasmodic coughing. He had the barrel-chest and lifted shoulders of the emphysema sufferer. His skin was greyish yellow, his hair sparse, his nose as formless as a lump of clay. His eyes were bulbous behind thick lenses.

At his side was a Bird respirator and a cylinder of oxygen, for breathing had become a major difficulty for him. The hands gripping the arms of his chair were blue-nailed, as if the process of decomposition had already set in.

The man called Luther sat down on a bentwood chair. His derisory gaze ranged round the room.

'Come down in the world, haven't you? Don't they give you a pension?'

The taunt did not disturb his host. 'It's quiet here. I have enough money to do my work.'

'What work? They trashed you! You don't belong in the Service any more.'

'De Klerk and his toadies may have gone soft, I haven't. I still have friends at the Shop. I can get information. Money if I need it.'

'Bitter-einders, all of you.' Luther spoke with contempt. 'There's no terrorists, any more, it's all forgive and forget. There's an amnesty, didn't you know?'

'Ersatz. It won't hold. It will be revised as soon as a new government is elected, and it won't indemnify hired killers. How many hits was it, Luther? Sixteen, seventeen? You're on the black list. You're as good as dead.'

Luther leaned back, smiling. 'What if I decide to take you with me?'

'Don't talk rubbish. I know how to protect myself. You should rather think of me as your friend, your ticket to freedom. Work for me. It won't be for long. My chest's kaput, the doctor gives me three months. When I'm dead you can go back to chasing cows – or better still, use the cash to start a new life. Plenty of scope for your sort in South America.'

Luther turned his head away. He knew that anger was counter-productive, and forced himself to speak calmly.

'What are you trying to do?'

'Finish my job before I die.' The words had passionate sincerity. 'Know what I am? I'm a pit bull terrier, I never let go. For ten years I sent in reports. I warned them what was going on. I gave them facts, figures, names – big names, some of them – inside the system and out. Nothing was done. No official action. I did my duty and they humiliated me, kicked me out. Things are different now. I can make them listen to me.'

'How?'

The squat man chuckled, enjoying his moment.

'Wivern's alive.'

Luther's face, usually devoid of expression, creased suddenly.

'You're crazy.'

Indeed there was something disquieting in the squat man's

demeanour, something that suggested the abnormal. He watched Luther with malign glee. 'Can't admit that Wivern spoiled your score, can you? The one that got away! Well, he's alive. I've seen his trail; slug's trail. He's using his old agents, drops, safe houses. Know what? I'm going to put out slug bait. Prove I was right all along. Then they can crawl to me.'

Luther made a derisive sound. 'You going after him in that wheelchair?'

'No, no. I'll leave the action to Scallan.'

Luther laughed outright. 'Scallan won't help you. He won't go to the Press, he won't make any public fuss.'

'So you're an expert on Scallan, now?'

'I know him. A few years back, the Firm thought of taking him out. I was detailed to watch him, learn his patterns. I can tell you he protects his wife and family. He won't go public.'

'And I can tell you, if he thinks there's a chance Wivern's alive, he won't rest till he finds him.'

'Then he's off his fucking head, like you.'

The squat man giggled. 'Sticks and stones may break my bones but words will never hurt me.' He groped under his pillows and drew out an envelope, tossed it to Luther.

'Better count it, ou boet.'

Luther put the envelope in his pocket and stood up. 'What if I report what you said?'

'Report to whom, my dear? These days, it's so hard to tell one's friends from one's enemies. Why run your neck into a noose? Take my advice, be patient. Stay close to the phone. I'll call you when I need you.' Leaning back, he closed his eyes. 'Shut the door after you, please.'

Back in his car, Luther counted the notes, then sat holding them for some minutes.

The old man was kranksinning, stone mad. Everyone

knew that. But the money was real, and he needed it. He needed it to get away, before the witch hunt started.

He put the money in his wallet, and drove quietly towards the S12 freeway.

6

On Saturday morning Todd rose at four. In frosty darkness he stowed his gear in the station-wagon, and drove to the garage for petrol. Coming back, he collected the sandwiches and coffee he'd ordered from the office the night before.

At ten minutes to five Thulani appeared at the front steps of the hotel, carrying a duffel bag and a large sack of vegetables. As Todd moved down to join him, a police van drew up on the verge of the road, and Sergeant Ritter climbed out.

'Morning, Mr Scallan. You up early.'

'You're up late, Sergeant.'

'Ja, that's so. Going to Jo'burg, are you?'

'Yes, to see my family.'

Ritter's eyes, red-rimmed, surveyed Thulani. 'Taking him with you?'

'Yes. He's going to visit his uncle in Soweto.'

Ritter nodded. He was a stocky man of thirty-five with frizzy black hair cropped close, and a broken nose. Black stubble showed on his jowls. He moved towards Todd.

'I hear the two of you was working up on Crown Ridge, Friday a week ago.'

'Yes. Who told you?'

Ritter frowned, debating with himself. Then he said, 'I was talking to Mr Kramer, over at Riverbend. He said he saw you. I wonder, did you see anyone else while you were up there? Any blacks?'

44

'No. The area seemed deserted. Are you thinking of the Smedley murders?'

'That's right.' Ritter glanced at Thulani, and said in Zulu, 'You boy, what's your name?'

'Msomi. Thulani Msomi.' Thulani spoke quietly, gauging the policeman's mood.

'Did you see anyone from this place?' Stubby fingers indicated the village and surrounding kraals.

'No, sir.'

The sergeant made an impatient sound. 'Put your things in the car,' he directed. As Thulani moved his bag and sack towards the station-wagon, Ritter turned back to Todd.

'They never see anything, hear anything. They stick together.'

'Would you inform against a neighbour, if you were black?'

'Prob'ly I wouldn't. All the same, if we can't get anyone to testify, we can't make a charge stick.'

Todd watched him. 'Who was the other man with Kramer, on the ridge?'

'Other man?'

'He told me he was an agent for fertilizer. Didn't Mr Kramer mention him?'

'No. What's he look like?'

'Smallish, slight build. Brown hair, sallow skin, brown eyes, dressed in jeans and a floral shirt, with fancy boots.'

Ritter's brow cleared. 'Ach, that's Clem Finlay. What'd he have to say?'

'He said the police were looking for two labourers Mr Smedley fired; but in the pub that night, opinion was they didn't do the killing.'

Ritter grunted. 'Talk's easy.'

'Do you know what guns the killers used?'

'An AK47, and an American parabellum.'

45

'Where would farm labourers get arms like that?'

Ritter's teeth glinted in the travesty of a smile. 'Mr Scallan, if I knew that, I'd be a colonel in Durban instead of a bloody sergeant in this dorp.' He stepped back. 'Ta for the chat, eh? I expect you want to be on your way.'

Todd climbed into the car and opened the passenger door for Thulani. As he made a U-turn, he saw that Ritter was walking up the road towards the store.

Driving along the feed road that would bring them to the northern freeway, he said to Thulani, 'What do the people say about Sergeant Ritter? Is he all right?'

To anyone but Todd, Thulani would have returned an evasive answer. Now he said, 'They wonder why he's here.'

'What do you mean?'

'Ritter is a clever man. He can speak English, Afrikaans, Zulu. He has medals.' Thulani brushed thin fingers across his chest. 'My grandfather says a policeman is sent to this place because he is very good or very bad. I don't know which Ritter is.'

Todd sighed. He had chosen to work in conservation because it took him into the deep country. A few years ago there'd been peace in these hills. A few heads broken after a wedding, perhaps, a few cattle stolen; nothing more. Now the violence had spread from the cities like a contagion. There was no refuge from it, anywhere.

He handed the pack of sandwiches to Thulani and asked him to pour coffee into two paper cups. They ate and drank in silence. By the time the sun was above the eastern plain, they were well on their way north.

They reached the Transvaal soon after eight o'clock. Todd took the ring road south of Johannesburg, heading west towards Soweto; but when he tried to turn on to the subsidiary road that led to Diepkloof, he found it blocked by a cordon of policemen and soldiers. A sergeant in battledress rapped on the window of the car, and Todd wound it down.

'You can't go in, sir,' the man said. 'There's trouble.' Listening, Todd could hear the distant rattle of automatic rifle fire. The haze of smoke that always cloaked the vast sprawl of Soweto was pierced in several places by towers of flame.

Thulani said anxiously, 'I can walk, Todd, I can find the house.'

'No, it's too dangerous. You're coming home with me. Joe will be delighted.' He gave the boy a reassuring smile. 'We'll phone your uncle and make a plan.'

He drove north through Westgate to the lush properties of Parktown. His father's house stood on an acre of ground, a graceful building with the curved planes and deep porches of the nineteen-thirties. The high white wall that encircled the property had been crowned with razor spikes, and there were electronic controls at the gate. Todd took a metal disc from his pocket and pressed it into the slot of the machine. The gate swung open and he drove to the house, stopping at the side entrance that led directly to his apartment.

He rang the doorbell and went back to the car to help Thulani unload. Before they were done, the door was flung open and a woman and child hurried out.

Aline Scallan was small, dark and stockily built, her only points of beauty a fine matt skin and magnificent taffy-brown eyes. Her son, who was eight years old, resembled her except that where she gave an impression of wiry strength, he looked frail.

He rushed to Todd, who swung him up in his arms, saying, 'How are you, sport? You OK?'

'I'm fine.' Joe was already wriggling to be set down, swinging an arm towards Thulani. 'Hey, man! Gimme five!'

The two boys struck palms, and raced off towards the house.

Todd kissed Aline. 'There's rioting in Diepkloof,' he said. 'Thulani may have to stay with us.'

She nodded, her eyes scanning his face. 'You look tired.'

'It's been a busy week.'

She continued to watch him anxiously. 'Those people who were murdered. Was it near where you work?'

'Not really. They farmed on the other side of the village.'

'I worry about you, out there on the veld.'

'I'm careful.'

'You should be armed.'

Todd's expression hardened. 'You know I hate guns. Carrying one invites others to do the same. The Smedleys were probably killed for their guns.'

'It's horrible. Old people like that being killed. Who could do such a thing?'

'The police will get them in due course.'

'They said on the radio they were sending in extra men.'

'Don't worry about it.' Todd put an arm round her shoulders and walked her to the house. 'What time's Ma's plane?'

'At three. We're all going to meet it.'

'Dad too?'

'Yes. He phoned from his office early this morning. He wants to invite some people to dinner. Something urgent he has to discuss.'

Todd frowned. 'Did he tell you their names?'

'A Dr Myra Rosenwert, and a Mr Max Duma. Do you know them?'

'Myra Rosenwert was a political scientist at the university. She retired a few years ago. Looks like a witch, but talks amusingly. Dad called her in as an expert witness in several of his cases. She taught Dave, and thought a lot of him. She wrote a book last year.'

'What about?'

48

They entered the hall. Todd said, 'Ma will freak when she arrives home to find Dad's arranged a dinner party.'

'She won't have to fuss. Philomela's doing roast lamb and I've made soup and some puddings. What was the book about?'

Todd did not meet her eyes. 'Political assassination,' he said.

7

On Saturday morning Neil Scallan called at the office of Major Tiaan Joubert, at the Computer Centre of the South African Police.

Despite his rank, Joubert was little heard of at head-quarters, and quite unknown to the ordinary citizens of Johannesburg. He belonged to that select breed of law-enforcement officers who are recruited for their special skills . . . in his case, a knowledge of the technology involved in modern criminal investigation.

Just thirty-seven years old, sandy-haired and bespectacled, Joubert looked more like a research clinician than a police-man. He disliked regulations, liked jazz and Greek food, and for exercise played badminton or squash. He possessed a master's degree in computer science, and had spent time in the USA, Britain and Germany studying police methods. He held strong views on how to combat the rising tide of viol-ence in the world.

It was for this last reason that he had been detailed to assist COVISA . . . the Committee on Violence in South Africa, appointed by the Transitional Executive Committee to map past and present political violence, and to draft legislation for a peaceful future. A law on amnesty was an important item on the Committee's agenda.

Neil had begun by distrusting Joubert, believing him to be nothing more than the latest model of security policeman, a species he hated. As time went on he was forced to concede that Joubert was not only clever, but dedicated to his job. He avoided the red tape and double-speak of officialdom, and he had his own ways of leaching information from the monolithic records of the police department.

During the initial meetings of COVISA, Joubert had remained a silent observer. Then one day he announced that he had set up a reference room for the group's use. 'It will give you an idea of what's available on computer,' he said, 'and show how one can draw on the data banks.'

The group visited Joubert's headquarters next day. Situated in a run-down western suburb, the building looked from the outside like a dowdy apartment block. The computer was housed in a specially constructed basement where it was protected from damp, dirt, unauthorized prying and physical attack. Joubert had set up his reference room in an office on the ground floor.

The left-hand wall of the room was covered with charts headed 'non-political violence', which listed every conceivable crime from minor assault to arson, kidnapping and murder. Against each item was an index number.

The right-hand wall was devoted to politically linked violence, and here the charts provided an appalling summary of the mayhem of the past decade: death in detention and the assassination of policemen; necklacings, riots, brutality at kangaroo-courts and in police stations; tribal vendettas, deaths at political rallies and funerals; clashes between security forces and schoolchildren; killing rampages in shopping malls and on suburban trains; a catechism of political unrest and social anarchy.

One chart listed the names of political activists who had been injured or who had died in violent circumstances. Among the names was David Malcolm Scallan.

A third wall was given over to computer terminals and a

large display screen. Joubert showed how it was possible to project on to this screen information relating to any item listed on the charts.

'Let's say you want to know about the attacks on train commuters' – he was tapping keys as he spoke – 'your reference is the index number on the chart. Within that reference, I or one of my staff can call up the data you need. For instance, you want to know where the attacks occurred' – a map appeared on the screen, starred with points. 'All on the Rand, as you can see, and all within a certain radius of the Pretoria–Johannesburg area.' The map vanished and lists rolled down the screen. 'This gives the breakdown of the attacks . . . date, number of casualties, witnesses, and so on.' More tapping. 'I'm working on a system of cross-reference. For instance, in Inanda in Natal, it's interesting to find that certain warlords who are extorting protection money from the residents, which we term a non-political crime, also pay the youngsters in the political cadres to destroy houses or livestock belonging to those who don't pay up. So you get an overlap of political and non-political violence. You find the same thing in America, where the Mafia or the drug cartels link up with corrupt politicians to achieve a common goal.' Joubert warmed to his theme.

'One of our difficulties in South Africa is this intertwining of the roots of violence. It's hard to say how much springs from ordinary criminal activities; how much comes from the historical animosity between tribes or factions; how much is part of the power struggle between the government, the ANC, Inkatha, and other political organizations, black or white; how much is socio-economic, the outcome of poverty, ignorance and disease.

'I am trying to amass in this centre as much information relating to the violence as I can; and I'm trying to discover if there are common factors in the main unrest areas across the country. Do certain names or groups recur? What is their *modus operandi*? And so on.

'The data is at your disposal. You can call on us at any time, the centre is manned twenty-four hours a day. I'll arrange the necessary security passes.'

As the group was leaving, Neil hung back.

'Major Joubert, what do you have on hit squads?'

Joubert touched a finger to his spectacles. 'So far, no one has admitted they exist . . . that is, if you're talking about political assassination.'

'What about covert State operations? The CCB?'

'Our records include newspaper reports, reports from unrest monitors, statements by people alleged to have knowledge of such squads. I file whatever is available to me.'

'I imagine there's a great deal that's not?'

'I don't have access to classified information.'

'So really, all you can provide is a rehash of what one's already seen in the Press? Shadows. No substance.'

'Oh no, I have more than that, Mr Scallan. After all, if a man sees the shadow of a vulture, he assumes the bird is close.' His bland gaze met Neil's. 'He keeps his eyes and ears open.'

After that, Neil made a point of visiting the centre once a week to talk to Joubert.

Joubert looked up in surprise as Neil entered his office.

'Mr Scallan . . . I didn't expect you until Monday.'

'I'm sorry to arrive without an appointment. It was a matter I didn't wish to discuss on the telephone.'

Joubert gave his owlish blink, then rose and went to close the door. Coming back, he said, 'What is it? How can I help?'

Neil settled himself in a chair. 'It's about my son, David Malcolm Scallan. His name's on that chart.'

Joubert nodded wordlessly.

'How much do you know about him?' Neil demanded.

Joubert considered. 'My first task when I was appointed to COVISA was to obtain security clearance for all its proposed members, including yourself. I know a good deal about you, sir, and something about every member of your family. Your son David is on record with the security police as having been an activist who helped political refugees to leave South Africa. He associated with known Communists.'

'In those days, anyone who opposed apartheid was dubbed a Communist.'

Joubert shrugged. 'Some of them were.'

'Most were not.' Neil folded his hands on the desk, kept his voice level. 'You know how David died?'

'Yes. He was doing a holiday job at the offices of the South African Council of Churches. At eight a.m. he left his digs to go to work. He walked to his car, which was parked on a vacant lot a short way down the street. He got into the car. When he turned on the ignition, it blew up. Your son was killed outright.'

'He was blown to pieces,' Neil said. 'There was nothing left to bury. Your estimate . . . the police estimate . . . was that the bomb contained fifteen pounds of explosive.'

Joubert said nothing, and Neil continued, 'The police never discovered who was responsible for planting the bomb. David's murder remains unsolved, like so many others. I tried for years to secure justice for my son. There were leads that were never followed up, people who were never questioned.'

'The record states that you talked of a police cover-up.'

'I did, yes. I knew there were hit squads. I'd seen their work before. I wanted those people dragged out of their rat holes, punished, eliminated from society. I spoke to everyone I thought could help, up to and including the minister.' Neil ran a hand over his eyes. 'It was before your time. In those

54

days it wasn't possible to talk to a police officer the way I'm talking to you.'

'Things are different now.'

'Are they? Yesterday afternoon, at the Trade Centre, I took a call from someone calling himself George Wivern. Does the name mean anything to you?'

Joubert frowned. 'Captain Wivern was one of the officers who investigated your son's death.'

'Right.'

'But he's dead, sir. He was killed in Angola when a petrol dump blew up. His remains were sent back to this country for burial.'

'His ashes. Not his body.'

'His unit officer provided identification. The official inquiry recorded the facts of his death. Wivern is dead!' As Neil preserved a stubborn silence, Joubert said in exasperation, 'Someone else spoke to you yesterday.'

'In Wivern's voice. He came from Malmesbury in the Cape. He had the Malmesbury burr.'

'That can be imitated.'

Meeting Joubert's stare, Neil shrugged. 'I know. I'm not hysterical. I admit there was a lot of background noise.'

'So. What did he ask you to do?'

'He told me to go to the information desk at Jan Smuts Airport. I did so, and was given this.' Neil leaned across the desk to give Joubert the envelope and its contents.

Joubert studied the slip of paper, then said sharply, 'Do you have a witness to receiving this?'

'I do. Mr Gordon Leveret was with me when I took the call, and he accompanied me to the airport.'

'We'll have it tested for prints. Probably a waste of time.' Joubert read aloud. 'Colonel Hermanus Silbermann, R24,000. Sergeant Lemuel Moroka, R7,500. Mr Mark

Alberton, R17,000.' He looked at Neil. 'What are you suggesting, Mr Scallan?'

'Those men were Wivern's associates. Silbermann and Moroka were on the team investigating David's murder. Alberton was a police informer on the university campus. The figures represent the money they were paid.'

'That's nothing but a wild guess! The money could relate to investments, pension funds, anything.'

'It relates to payments made to a hit squad.'

'Mr Scallan, you're under a lot of strain, but as a lawyer you know that in a court of law . . .'

'I'm not thinking of a court of law. I'm remembering the murder of my boy. I want the case reopened.'

'There's no chance of that, sir.'

'That paper confirms what I've always believed, that Wivern killed David. I want it brought to the attention of the authorities.'

'They'll laugh at it, sir, they'll say it's a hoax. Some wimp trying to make a monkey out of you.'

'If I'm the victim of a hoax, I want that investigated. As a member of COVISA I'm entitled to some protection from malicious attack. I wish to lodge a formal complaint.'

Joubert sighed. 'I'll arrange for someone to take your statement.'

'I'd prefer you to handle it yourself, Major. I don't want any publicity. My wife and family must not be disturbed in any way.'

'Very well.' Joubert reached for a pen. 'But believe me, Mr Scallan, it's chasing shadows. George Wivern is dead and gone.'

'I never believed that,' Neil said flatly. 'Why should I start now?'

8

Before he sat down to breakfast, Todd put through a call to Thulani's uncle. Philip Msomi was at home. He sounded tired and worried.

'Mr Scallan, I meant to call you from the bank, but I can't go in this morning. They're burning and looting only a few streets from here. My wife and I must stay and look after the house. I am anxious to see Thulani, but it's not safe for him to come to Diepkloof, and I can't leave.'

'Don't worry, Thulani can stay with us. If things settle down tomorrow, we'll make a plan to meet. If not, there'll be other weekends.' He cut short Msomi's thanks and handed the receiver to Thulani.

At the end of that conversation, Philip said, 'Tell my father that I have read his letter and will do as he asks. Tell him I'm hopeful the bank will help in the matter he mentioned.'

'Yes, Uncle.' Thulani felt pleased, for he was sure the letter concerned his own future.

'And Thulani, hear what I'm telling you. Don't give Mr Scallan any trouble. He's been very good to you. You should think how to make him a nice present.'

During the course of the day Thulani gave a great deal of thought to what such a present might be; but looking at Todd's apartment with its fine furniture, and the big garden with its swimming pool and tennis court, he was at a loss to think of anything Todd could need.

*

Meg's plane from New York was delayed by bad weather. She arrived twenty minutes late and in a fractious mood. Having embraced Todd and Aline, and kissed the air along- side Neil's cheek, she said, 'Where's Joe? Why isn't he here? I was looking forward to seeing him.'

'He's home with a friend,' Aline said. 'Todd brought Thulani Msomi with him, and he can't go to his uncle because there's rioting in Diepkloof.'

'All that again! I must say it was heaven to be in America, in a civilized ambience. Everyone was so kind. Ronelle's a marvellous hostess, she's really helped to put Alistair where he is.' She turned to Neil. 'Did you bring both cars? With all my luggage . . .'

'I brought Todd's station-wagon, there'll be plenty of room.' He saw the restless movements of Meg's hands, fid- dling with her purse and the scarf at her throat. There were dark shadows under her eyes. In the car she chattered brightly about her holiday; but when she learned there were to be guests for dinner, her face became thunderous.

'My first night home, for God's sake! Really, you might have had some consideration . . .'

'Aline's seen to everything, my dear, you won't have to bother . . .'

'I'm not worried about the arrangements! I should think I know how to fix a dinner party by now! But I'm tired out. I wanted it to be just family tonight; I wanted to talk to Todd. Who are these guests, anyway?'

'Myra Rosenwert and Max Duma.' Meeting Meg's glare, Neil said placatingly, 'It's business, you see.'

'On my first night? Won't tomorrow do?'

'No. It's very urgent. We won't trouble you. After dinner we'll go off and do our talking, and you can have the family to yourself.'

Aline intervened with a story of Joe's performance in the

school concert, and Meg appeared to be mollified, but Neil was not deceived. He knew she was coming off the high provided by her visit to the States. It would take very little to plunge her into depression.

There was no way he could tell her about the message he had received from the man calling himself Wivern.

The two people Neil had invited to dine were friends of long-standing; political animals, well-versed in the lore of the apartheid years, able to contemplate its nightmare landscapes without illusion or surprise. He needed such company now, and hearing the note of anxiety in his voice, they both accepted his last-minute invitation without demur.

Myra Rosenwert was the first to arrive, driving up to the door in her old black limousine that looked like a hearse, climbing out with a flurry of black draperies, resting gnarled fingers on Neil's lapel as he bent to kiss her.

Myra had had polio as a child and it had left her with a slight limp. That, with the sharpness of her nose and chin, gave her a decidedly witchlike appearance, something she exploited to the full. Her sole beauty was her voice, which was rich, warm and flexible. She had been widowed twice and was a wealthy woman, but she had chosen to go on teaching at the university as long as the rules allowed. She still gave occasional lectures by invitation.

She leaned heavily on Neil's arm as they climbed the steps to the front door. In the drawing room she greeted Meg with circumspection, visibly curbed the force of her personality, and talked of commonplace things. Meg, who was no fool, knew she was being humoured and resented it. She began to question Myra about the results of an opinion poll that had appeared in the evening paper, asking if she agreed with its findings.

Myra took the glass of dry sherry Neil was holding out to

her, and sipped before replying. 'It's difficult to disagree with any properly conducted poll,' she said. 'After all, polls only tell us what we already know . . . or should know if we used our five senses. What this country needs is the sixth sense. We need prophets.'

'I would have said prophecy was the gift of God.'

'Oh, it is, if you mean major powers. Only a very special vision can bridge the centuries, as did Isaiah and Ezekiel. I'm speaking of the ability to see a year or two into the future.'

Meg's face seemed to shrink. 'I don't want to be able to see the future. It's hard enough to bear the present.'

'Sufficient unto the day is the evil thereof?' Myra turned as Todd approached, stretching out a hand to him. 'Aha. How good to see you. How are the grasses?'

'Blooming.' Todd liked Myra and smiled warmly at her, so that she caught a momentary likeness to his father. Momentary only, because Todd would never possess his father's powerful presence, his aggressive charm. She patted Todd's wrist. 'I'm glad,' she said. 'I was surprised when I heard the Institute was to do the survey. I thought you'd been turned down.'

'We had, because they wanted to push ahead with the dam. The recession put an end to that, so they're making the best of a bad job, and doing what they can to save flora and fauna.'

'Will there be enough time?' Myra knew that to make a complete survey of an area took months, if not years.

'I hope so. Even if the recession lifts, they won't start work on the dam all that fast. The workers were all paid off, which wasn't a bad thing. Most of the Muller's Drift blacks support Inkatha, and the imported labour was predominantly ANC. There were several run-ins, and two men were killed.'

Meg said in alarm, 'You never told me there was fighting.'

'It was before I got there,' Todd assured her. 'All over bar the shouting.'

She made a fretful movement. 'I wish you'd come home, Todd. I wish you'd find a proper job. The Institute pays you next to nothing. You could find something much more suitable.'

'This suits me, Ma.'

'Is there a doctorate in it?' Myra asked.

'Shouldn't think so.'

The sound of a booming voice in the hall prevented further discussion. Moments later Max Duma entered the room. Large of face and frame, he exuded a raffish *bonhomie*. He was a sharp businessman. During his twenty years in exile, he had raised massive sums for the ANC. His enemies maintained that he had in the process handsomely feathered his own nest, but Neil knew this to be untrue. The suit might be tailor-made, but the man inside it was homespun. The grassroots people loved Duma, and trusted him implicitly.

Meg detested him; not for himself, but because to her he represented the political force that was bent on overturning her world. It infuriated her that Neil should invite Duma to dinner, tonight of all nights, but for Todd's sake she greeted him with courtesy. When the new South Africa dawned, men like Duma might be disposed to find a safe place in it for Todd.

At dinner the talk turned inevitably to politics, and in particular to the forthcoming elections. Myra chose to be provocative, needling Duma.

'In principle I support the ANC,' she said, 'but I have to tell you I'm heartily sick of sabre-rattling and mass demos. It's time you stopped behaving like a resistance movement and began to behave like a government-in-waiting . . . since that's what you claim to be.'

Duma spread his hands. 'We have to use the tools at our

disposal. We can't rely on Saatchi and Saatchi, like the Nats. Many of our voters can't read, they don't have television. We communicate with them through mass action.'

'You drum up emotions, but is that really communication?' Her black eyes challenged him. 'Do any of the political leaders in this benighted country have real control over their followers at grassroots? Are any of them prepared to drop the power-play that's killing thousands of ordinary people? How do they plan to end the violence?'

The talk became general, the familiar clichés tossed about like ping-pong balls. Social upliftment. Constitutional reform. Economic growth.

Neil rose to refill the wine glasses. Coming back to his place, he turned to Todd, silent at the far end of the table.

'What do you think, Todd?'

'I think there are too many guns.' As everyone fixed their eyes on him, Todd shifted uneasily. 'Peasants in mud huts are being cut down by AK47s. Whoever killed the Smedleys used an AK47 and a parabellum.'

'Who are the Smedleys?'

'An elderly couple who farmed at Muller's Drift. Two of your "ordinary people".'

'I suppose they were robbed?' Meg's voice was shrill. 'Old folk are soft targets these days.'

'They had nothing worth stealing, Ma.'

'That wouldn't stop that kind of thug.'

Duma spoke. 'What do the police say?'

'Sergeant Ritter's looking for two labourers Smedley fired.'

'Then that's the answer.' Meg put both hands on the table. 'Now can we talk about something else? I'm sick of hearing about killing.' Her mouth was pinched as if she might break into tears, and Myra said quickly,

'You're quite right, one becomes morbid. Tell us about your trip, Meg. How's Alistair?'

Meg beamed. They plied her with questions, stepping delicately past the dangerous topics as people avoid the chalked outline of a body at a murder scene.

9

'You've been very quiet, Neil,' Myra said. 'What's on your mind?'

The meal was over and Meg and Aline had gone to join Joe and Thulani in the television room.

'I received a telephone call yesterday,' Neil said, 'from a man calling himself George Wivern.'

Duma checked in the act of lighting a cheroot. 'Wivern?! He's . . .'

'. . . officially dead. I know. The caller told me to go to the airport, where I collected this.'

He handed a photostat of the note to Myra. She read it quickly, her face puckered.

'Mark Alberton was a campus spy. And the other two, Silbermann and Moroka . . .'

'. . . worked with Wivern on the investigation into David's murder.' Neil passed the note to Duma, whose reaction was immediate.

'It's what you said all along: David was killed by a hit squad. The same people who investigated the case, planted the bomb.'

'And Alberton?'

'Kept David under observation, gave warning of his movements, shopped him to the cops.' Duma paused. 'One thing, though . . .'

'What?'

'Don't take offence. The payment seems too high. It's not a question of what your son was worth to you, but what he was worth to them. A student, helping other students to skip the country? He can't have been so high on their hit list.'

Duma made as if to pass the note to Todd, who shook his head.

'I've seen it. I think it's a hoax. Some nut wants to hurt us. Hurt Dad.'

Myra looked at Neil. 'That's probably true, Neil.'

'Probably. It may be a malicious hoax. Someone may be trying to unsettle me, divert me from my work on COVISA. That's what reason tells me. My gut tells me different.'

'Tell us exactly what happened.'

He told them the full story. At the end, Duma said, 'Did you go to the police?'

'I spoke to Joubert this morning. I have to be careful. I don't want any publicity. Meg mustn't know about it. On the other hand, I can't just ignore it. I'll make discreet inquiries. That's where you two can help. Myra, you knew Mark Alberton. You taught him, didn't you?'

'Yes, he was in Pol Phil for a year, with David.'

'Did you suspect at that time he was an informer?'

'No. Nor can I believe he was paid seventeen thousand rand. He was a lightweight; I would have said a nothing. He infiltrated the left-wing organizations on campus, but he was regarded by them as something of a joke. I doubt whether he ever had much to sell to his pay-masters. Did you ever meet him?'

'Dave never brought him home. I did meet him on one occasion. It was at a varsity function. He came up and spoke to me, said he was interested in becoming a lawyer. After he was exposed . . . after David's death . . . I had no wish to see him.'

Myra narrowed her eyes, remembering. 'He was so nondescript. Older than the average student, about twenty-five or -six. Tall, rather stooped, with brown hair that fell across his forehead. Brown eyes . . . he wore glasses . . . pale skin. What would be called a nerd, nowadays. A hard worker, but I doubt he'd ever have made more than a second-class degree.' Myra frowned. 'I think he was looking for more than money when he became an informer. He wanted recognition. Approval.'

'What about his family?'

'He was an only child. His parents came to talk to me when the Press broke the story that he was a campus spy. They were plain folk, very conventional, very distressed. The father owned a garage in Brakpan. The mother was Portuguese, plump and sallow with kirby-grips in her hair. She didn't say much, just cried. Mr Alberton said Mark believed he was doing his patriotic duty, "flushing out subversives". Really, it was a pathetic interview. I advised them to withdraw Mark from the university. They did so. Later, he was expelled.'

'You say he was a nothing,' Duma said, 'but he's on that list for seventeen thousand rand. That's a lot of money. Maybe the nerd was smarter than he looked.'

'Maybe.' Myra sounded unconvinced.

'What happened to him after he was kicked out of Wits?'

'The family emigrated to Australia. Mr Alberton was born there, so had Australian citizenship. About a year after they left, I had a letter from him saying they were happily settled and making a new life.'

'Can you give me the address in Australia?'

'I'll see if I can find it.'

Duma asked the question everyone else had asked.

'If Wivern's alive, why would he come to you?'

'I don't know. Perhaps he's looking for some kind of trade.'

'With you? Wasting his time, isn't he?'

A muscle jerked at Neil's jawline. 'He can apply for indemnity like anyone else. A court will decide, not me.'

'Yeah, right. In the mean time, what can I do for you?'

'Find out what you can about the men on the list. Silbermann, Moroka, Alberton and, of course, Wivern himself. Find out if they were attached to the SB or any covert operation. See if you can discover who was their controller. I want the spider at the centre of the web.'

Later, as Myra climbed into her motorcar, she said to Neil, 'Todd mentioned a Sergeant Ritter. The name rings a bell with me, something to do with guns. Would you like me to check my files?'

'Please.' He gave her a twisted smile. 'I know I'm clutching at straws.'

'Aren't we all?' Looking at him she felt a stab of pure rage. Whoever was doing this to him should be nailed for it. 'I'll let you know,' she said, and with a final saurian grin, closed the car door and drove away towards the gate.

Snug in the television room, armed with trays of food, a video of *Rocky V* unwatched on the screen, Thulani and Joe discussed important issues: boxing, soccer, athletics, and the use of steroids in sport.

'I heard Dad say it's stupid to use them,' Joe said. 'If you get caught they won't let you play, and even if you don't, it kills your sex drive.'

Thulani chuckled. 'I'll tell my mother, she can feed it to Gwala.' He leaned forward and switched off the television set. 'Joe, can you tell me, what can I give your father that will please him very much?'

Joe shook his head. 'He wouldn't want anything.'

'But I wish to make a present. He got me into the college, he gives me holiday jobs.'

'Well, he says you're worth it. You tell him the names of plants and stuff like that. You don't need to give him anything.'

Thulani was silent for a while, then got to his feet and studied the rows of reference books that lined the walls. Finding what he was looking for, he brought it back to the sofa. He turned to the index section and ran a finger down the columns.

'*Woodii*,' he said. 'Do you know about that?'

'Sure. It's a cycad. You can't give him that, Thulani. They're very rare.'

Thulani wasn't listening. He flipped through the pages of the book until he reached the section headed *Encephalartos woodii*. There were photographs of a fully grown plant, and scale drawings of the long arched rachis, the prickly leaflets, the male cone, smooth and yellow-orange.

'Do you have tracing paper?' he asked. 'I want to copy these.'

Joe went to the kitchen and returned with some grease-proof paper. 'You can see through this,' he said. 'Don't press too hard, you'll make lines on the page.'

Thulani made careful tracings of the illustrations. That done, he went through the book page by page. At one point he stopped and read the text several times. Joe, becoming bored with this studiousness, asked what he was doing.

Thulani turned a shining face towards him.

'Nothing,' he said. 'A surprise. I don't know if it will work.' He folded his sketches and placed them carefully in the inside pocket of his blazer.

10

On Sunday morning Todd and Aline visited the art gallery that was exhibiting several of her pictures. Neil went with them, planning to leave early and call at Anna's flat; but while he was still at the gallery, Anna herself walked in. Neil went to meet her.

'I didn't get a chance to call you,' he said. 'Meg's back, and Max and Myra came to dinner last night.'

She nodded, cutting him short. 'I found an old file in the law department stackroom,' she said. 'It was mostly newspaper clippings, nothing you haven't seen; but there was this.'

She handed Neil a snapshot. It showed a big man moving up a flight of stone steps. At the foot of the steps was a black limousine. A passenger in the back seat leaned towards the window, apparently watching the man on the steps. His features were in shadow. Turning the picture over, Neil saw the pencilled words 'Outside the court, inquest on David Scallan.'

'You remember, the police wouldn't allow photographs that day?' Anna said. 'One of David's friends must have managed to get this shot, and put it in the file.'

'The man on the steps is Silbermann,' Neil said.

'And the other?'

'It's impossible to tell.'

'The quality of the print could be enhanced. I know

69

someone who could do it. He's a palaeontologist at the university. He has great photographic equipment.'

'Can he keep his mouth shut?'

Anna said coldly, 'Yes, if I ask him to.'

'What's his name?'

'Professor Ernie Bryceland.'

Neil handed back the snapshot. 'I suppose it's worth a try.'

Todd and Aline came up to them, Aline wreathed in smiles. 'I sold all five,' she said happily. 'Hi, Anna, how are you?' She sensed the tension in the air and the smile faded. 'What's wrong, what's happened?'

'Nothing,' Neil tried to sound cheerful, but she was not deceived.

'It's about that note, isn't it? Todd told me . . .'

Todd put an arm round her shoulders and said loudly, 'Leave it, Ally. It's only some mischief-maker.'

She pulled free of him. 'I'm not a child, Todd. If Neil needs help . . .'

'I don't want you involved.' Todd's voice had risen and people near them turned to stare. He went on more quietly, 'If you're finished here, I'd like to go home. I promised Joe.'

As Aline hesitated, Neil said quickly, 'Go along. If there's anything you can do, I'll let you know.'

She allowed Todd to lead her away. Neil stood stiffly, staring at the floor.

'He doesn't want to see you hurt,' Anna said.

Neil's head jerked up. He stared at her almost with hostility. She said again, 'None of us wants to see you hurt.'

He made an impatient sound, then shrugged. 'This Bryceland. Can we go and see him at once?'

She nodded, putting the snapshot in her handbag. 'I'll drive you,' she said.

*

Professor Bryceland lived in Yeoville. They found him in his work room, delicately cleaning a chip of fossilized bone. He was a gangling, slow-moving man with a shock of grey hair and heavy-lidded eyes; like something from the aeon of the giant lizards, Neil thought. There was the scar of an operation on his upper lip, and his voice had the furry quality of a repaired palate. He greeted Anna with affection, and listened in silence to her request.

Examining the snapshot, he screwed up his face. 'It's poor material,' he said. 'I don't know that magnification will help. I could mebbe enhance the quality, bring up the background.' He fixed Neil with a bright stare. 'How urgent is it, Mr Scallan?'

'To me, very urgent. Though I hate to interrupt your work . . .'

Bryceland grinned. 'That's kept a few millennia, it'll bide another day or two. Tomorrow? Will that do?'

'You're very kind.'

The old man waved the thanks aside and disappeared at once to his darkroom. As they made their way back to the car, Neil put an arm round Anna's shoulders.

'Thank you. I don't know what I'd do without you. Always on my side.'

She shook her head. Looking down at her, he saw that her eyes were full of tears.

The rest of the day passed with some semblance of normality. Meg kept to her bed, pleading jet-lag. Todd and Joe challenged Neil and Thulani at tennis, and were beaten by what Todd described as grubber tactics. Philip Msomi arrived at the house, spent an hour with his nephew, asked Todd to assure his father he would do all he could to help the boy, and drove off laden with Ellen's vegetables.

After supper Neil left the family listening to a Pavarotti

tape, and went to his study to put through a call to Rome. Almost before the phone had had time to ring, a woman's voice answered, speaking rapid Italian.

'Francesca?' Neil interrupted the flow. 'Neil Scallan.'

'Neil, my darling, how lovely to hear your voice!'

'And yours. Is Louis home?'

'No, alas, he's in Arabia, looking at oil wells. Can I help?'

'I'm afraid not, love, I need to speak to him. It's urgent, Fran. Can you get a message to him?'

'Not while he's out in the desert, but it's arranged he'll call me early tomorrow morning. What must I tell him?'

'Ask him to phone me, any time of the day or night. He has all my numbers.'

'I'll tell him.' Francesca rattled on for another six minutes, asking after Neil's family, insisting they must all come to Rome very soon, or better still to Greece; they could stay on the yacht, it would be so delightful. Her chatter sounded empty, but Neil knew that Francesca Baird was sound both in head and heart. She would see that Louis got the message.

Leaving the study, Neil encountered Todd on his way back from the kitchen with a sixpack of beer.

'Todd? . . . One moment . . .'

Todd halted, his expression mulish. 'What?'

'At dinner you mentioned a Sergeant Ritter. Myra recalls someone of that name – "something to do with guns", she said. Can you check up on the man? Find out where he served before Muller's Drift?'

'No, I can't. I told you, I don't want to be involved. The note's a hoax. You should ignore it.'

'How can I, when it might have a bearing on David?'

'David's dead, for Christ's sake, why don't you think of the living, for a change? You've driven Alistair and Ronelle away, isn't that enough? Do you want to put Ma in a mental home?'

72

'I've no intention of upsetting your mother.'

'How can you avoid it? How long before she finds out what you're doing? There'll be a load of muck in the papers, and she'll see it. She can't take any more pressure. As for Aline and Joe, they never even knew Dave, why should they be . . . ?'

'You knew him, God damn it, he was your brother! Your brother whose murderers are still walking around free, enjoying their blood money!'

Todd moved back a pace. 'Can't you get it through your skull, that note is a fake? There's no proof that Wivern and the rest killed Dave, no proof they were paid any money. The whole thing's a lie. I won't allow you to drag my wife and son into this, I won't allow Ally to help you, and I won't help you myself, do you understand?'

'I understand that as usual you choose to avoid unpleasant facts.'

'I don't choose to roll in them, like you.'

Todd swung on his heel and hurried into the living room. Through the open door Neil saw him rejoin the group by the fire, sit down close to Joe and put his arm about the boy's shoulders.

11

Todd and Thulani left Johannesburg at five on Monday morning, and were in Muller's Drift by eight. They found Hosea Msomi waiting for them at the hotel, very perturbed because he'd heard on the radio about the unrest in Diepkloof. Was it quiet now, he wanted to know, were Philip and his family all right, was the house safe?

Todd reassured him on these scores, passed on Philip's promise of support, and told Thulani to take the morning off and give his grandparents all the news.

He wanted time alone. The events of the past three days had released in him a storm of anxiety and fear.

It was impossible to explain, even to Aline, the effect that Wivern's name had on him. For months after David died, Wivern had persecuted every member of the Scallan family; questioning, insinuating that they had known of Dave's activities, keeping them under constant surveillance. Then, inexplicably, there had been a reversal of roles. Wivern became the hunted and his father the hunter, demanding justice and insisting that David was murdered by a hit squad. Quite suddenly, Wivern had left Johannesburg; sent to Angola, it was said, and killed there a few weeks later during an MPLA attack on his camp.

Hearing the news had been like waking from a nightmare; but now came the rumour that Wivern was alive. Todd refused to believe it. He would not allow that evil presence

74

to return. He would not help his father in another futile crusade. He would not be involved in any way. He had a wife and family to protect.

He had breakfast in the hotel, then set out for his work site, using the high contour road to the dam and avoiding Crown Ridge. He worked steadily all morning, collecting specimens and recording them in his notebook. It seemed to him that the countryside was unusually quiet. The stillness was eerie. Once he heard shots far down on the plain; but the silence rolled back, and he decided it was only a farmer after a rabbit.

At one o'clock he returned to the village, and spent the afternoon sorting specimens with Thulani. The boy seemed to sense his mood, and refrained from his usual chatter.

That night the public bar was almost empty. Looking around for Pieter du Toit, Todd saw him sitting at a table with Fanie Kramer, the farmer who'd confronted him on the ridge the week before. Todd went over to join them.

Kramer was friendly enough now, asking how Todd's work was going and how long he expected it to last.

'I'll take as much time as I can get,' Todd answered. 'At least until they resume work on the dam.'

'Then you got a long time,' said Kramer comfortably. 'They won't be in a hurry up there. No money in the kitty.'

The talk turned to the drought and the recession. At six, Kramer left, and his place was taken by Sergeant Ritter. Setting his tankard on the table, he grinned at Todd.

'Well, Mr Scallan, did you have a nice time in Joeys?'

'I enjoyed being with my family.'

'That's right.' Ritter took a pull at his lager. 'A man should have a wife and kids.'

'Are you married, Sergeant?'

'Divorced. My vrou walked out four years ago. Couldn't stand the life. A cop's a target these days.'

'Do you have any leads in the Smedley case?'

'Not yet. We'll get the buggers, sooner or later.' Ritter seemed to be making himself a promise. 'Meanwhile, you watch yourself up on the hills, Mr Scallan. You carry a gun?'

'No.'

'You should. Have to protect yourself out there. I can't do it, that's for sure.'

'They should send you more men.'

'Maybe they will, some day when the sun's not too hot and the moon's not too cold.'

Ritter finished his beer and departed. Todd turned to du Toit who sat brooding, turning his empty glass in his hands.

'Is he still looking for Smedley's labourers?'

'Ja. He must eliminate them as suspects, isn't that so? Doesn't mean he thinks they did it.'

'Hosea Msomi says the killers came from outside this area.'

'Could be.' Du Toit seemed about to say more, then changed his mind. He raised a hand to signal the waiter dozing by the door. 'What's yours?'

'Same again,' Todd said.

He thought the phrase expressed the nightmare round-about of his life.

That night Thulani told his grandmother that he'd thought of a present to give Mr Scallan.

'I hope it won't cost too much,' Ellen said. 'We're not rich, remember.'

'It won't cost anything.' Thulani was filled with glee. 'I must speak to Uncle Siphiwe about it.'

'You mean medicines? White people don't like Zulu medicines.'

Thulani shook his head, grinning. She saw that he meant to keep his secret and asked no more questions, merely cautioning him to be careful what he said to his great-uncle, as the old man was touchy these days.

Thulani set out directly after the evening meal.

Light showed in his uncle's house, but he approached the gate with circumspection, called out loudly to identify himself, and advanced only when he was bidden.

Siphiwe had been on the point of going to bed, and was disinclined to allow Thulani over the threshold; but when the boy explained the purpose of his visit, Siphiwe's professional interest was aroused. He invited Thulani in, set stools for them, and examined the tracings Thulani had brought.

'Yes,' he said after a while, 'I have seen such a plant, once. It was a long time ago, before the whites drove us from our land.'

'Mr Scallan says there are many that look like this one, but they are not this one.' Thulani pointed to his drawing. 'If my uncle will look carefully, he will see that the branches do not grow straight up, like spears, but bend over softly. There are thorns on these lowest small leaves, you see? And in the middle of the branch, the leaves are a little spiky.'

Siphiwe sniffed. 'Why does Mr Scallan like this plant? You cannot make any muti from it, you cannot eat it. It is quite useless.'

'It is valued because it is very rare, and because there's no female plant in the world.'

Siphiwe grunted, pointing to one of the sketches. 'What are these? They look like seeds, to me.'

'They are male seeds. They carry the powder that will make female seeds bear fruit, but if there's no female seed, the powder is wasted. Mr Scallan says a man can propagate new plants by taking suckers from the old one, but there will only be male plants.'

Siphiwe cackled, showing broken teeth. 'A life without women is no life, but does a plant care about such things?'

'As my uncle knows, all plants are valuable. They must be protected in order that we can continue to take what we need from them.'

'True, true.' Siphiwe pulled a horn box from his pocket, opened it, took a pinch of snuff and inhaled it noisily. 'If you find this woman-plant, will Mr Scallan pay you for it?'

'No. It will be my gift for him, because he found me a place at school, and buys my books and gives me work.'

Siphiwe scowled. 'We owe nothing to the Mlungu. It was not kindness that made them drive us from our good land and give us these rocky fields where little will grow!'

'That was long ago. Mr Scallan had nothing to do with it. He pays me well, and will help me go to university.'

'Ohé! And what will you do there?'

'Become a doctor of plants, like you.'

'Well . . . I will believe that when I see it.'

Thulani saw that his compliment had pleased the old man, and pressed the advantage. 'A plant like this would not grow in the full sun. It would prefer some shade, and good soil and water.'

'Fussy, like all women. Did Mr Scallan tell you these things?'

'No, it was in the book where I found these pictures.' As Siphiwe continued to look derisive, Thulani employed cunning. 'Of course, if it's only to be found on the Mlungu's land, then we can't reach it. We must forget all about it.'

'Their land?' Siphiwe was up in arms at once. 'It is not theirs, it is ours. They stole it, and some day they will have to return what they stole. Yes, sooner than they think. I have read the bones. They tell me a great storm is coming. It will sweep their farms and cattle away and carry them down to the sea! Ehe! That is truth. As for this plant you speak of, I have been thinking and I remember there were such. They had red seeds, packed close together like the teeth of a mealie.'

'Where, Uncle? Where have you seen the plant?'

'I can't remember. It was many years ago and, as I told you, the thing is useless, I didn't trouble my head with it.' Siphiwe sighed, searching the far reaches of memory. 'Ai, mai, it's not good to grow old. I can't remember.'

'Perhaps this will help.' Thulani spread one of his tracings flat on the floor. 'You see? Perhaps this one will guide us to the place, as the honey guide leads the badger to the honeycomb.'

12

Neil Scallan received two telephone calls early on Monday morning. The first was from Gordon Leveret, who sounded exhausted.

'Neil? This morning's meeting has been put back until three this afternoon, to accommodate the UN reps. Can you make that?'

'I'll do my best.'

'About Friday's . . . episode. I'm convinced it's a trick to make you pull out of COVISA. I beg you not to let yourself be sidelined. I've spent the whole weekend persuading people to stay with us. Negotiations are hanging by a thread . . .'

'I know. I'll make the meeting, don't worry.'

'Thanks, Neil. I appreciate it. See you later, then?'

The second call was from Myra Rosenwert, asking him to come to her home in Waverley. Arriving there, he found her in her study. She brandished a paper at him.

'The Albertons' addresses,' she said. 'Ronald Alberton lived with his wife Nesta and son Mark at Number seventeen Irvine Crescent, Booysens. They left in late 1983, after Mark was exposed as an informer. They went to Brisbane, Australia. About a year after they emigrated, Mr Alberton wrote to tell me they were safely settled. I'm afraid I destroyed his letter, but I kept the address. It's 201 Sandbar Road, Brisbane. They may have moved since then.'

Neil pocketed the paper. 'What about Ritter, did you find anything?'

'Not it my own records, but I thought of the unrest monitors in my old department. They lent me this file. It's not supposed to leave campus, so keep what you read to yourself.'

Neil carried the file to a chair and read the section Myra had marked. The first item was a cutting from the *Natal Herald*. The dateline was 20 September 1990.

Bus-stop Massacre in Vryheid

In the early hours of this morning two white men drove past a bus rank crowded with people on their way to work in central Vryheid, and sprayed them with bullets from AK47 rifles. Eleven blacks were killed and twenty-one injured. The victims included seven women and a two-year-old girl. Their names have not yet been released.

A police car patrolling in the vicinity gave chase to the assailants, finally forcing their car off the road three miles outside the city limits. The occupants of the car opened fire on the police, who retaliated, killing Mr Theodore Oppermann, a farmer in the district who was the self-confessed leader of the notorious White Torch movement.

The second assailant, later identified as Mr Charles Henley-Rooke, also a member of the movement, attempted to escape the scene on foot. He was pursued and

cornered by Sergeant C.A. Ritter of the
Vryheid SAP, and in the ensuing shoot-
out, Henley-Rooke was shot dead and Sergeant Ritter seriously wounded. He was
taken by helicopter to Gray's Hospital in
Maritzburg, where he underwent an
immediate operation. His condition is
reported to be serious but stable.

An editorial in next day's *Herald* commented:

The massacre of innocent black commuters
in Vryheid yesterday has shocked decent
citizens across the country. Coming at a
time when there is a move towards non-
racial talks, the attack could have inflamed
an already tense situation, triggering
reprisals from blacks and a trade boycott
that would further depress an already
reeling economy.
   The tragedy was redeemed to some
extent by the prompt action of the police in
pursuing and overpowering the assassins.
Sergeant C.A. Ritter cornered and brought
down an armed killer, himself incurring
grave injuries. It is to be hoped that his
courage and devotion to duty will be
suitably rewarded by the authorities.

There followed a number of reports from members of the
Unrest Monitors' Group who had interviewed surviving vic-
tims of the shooting, witnesses and policemen handling the
case.
Myra had starred one report in particular:
'I visited Ritter during his convalescence,' the monitor

wrote. 'I found him difficult to talk to. He's very much the policeman, unwilling to speak to outsiders even with the consent of his superiors, and mistrustful of anyone hailing from a liberal university. I think he told me the truth, though probably not the whole truth. He did admit that Henley-Rooke, the man he shot, was "a crazy who collected guns". He then muttered something I couldn't quite catch . . . it sounded like MacAllivan or Machiavellian; but when I asked him to repeat it, he changed the subject. I understand that Ritter has been commended by the Minister for Police, for his action after the massacre.'

Neil shook his head. ' "Machiavellian" seems a strange word for a man like Ritter to use. More likely he was naming some other person connected with the White Torch?' He laid the file aside. 'It could be the same Ritter . . . Todd never mentioned his first name . . . but if so, why was he transferred to a place like Muller's Drift? I wouldn't call that a fair reward for courage and devotion to duty!'

'Perhaps he was given a quiet post for his own sake,' Myra said, 'to let him recover from his wounds, or to keep him out of the way of white extremists?'

Neil did not reply. He was thinking that someone in authority might have preferred to have Ritter out of sight and out of mind.

Returning to the centre of the city, Neil decided his next move must be to try to trace the Alberton family. It was now 10.30 a.m. – late afternoon in Brisbane, Australia; a reasonable time to find people at home and awake.

Fearing his office phone might be tapped, he drove to his club. The privacy of a member was sacrosanct in that bastion of old-fashioned virtues.

He booked his call and waited in the booth in the foyer for the connection to be made. Presently a voice like a buzzsaw informed him there was no one called Alberton in the house; the people before had been Hirsch, and the people before them were Foley, and that was eight years ago, sorry mate.

Neil thanked the man and hung up. Although he had not expected to find the Albertons still at the address, he felt deflated. He tried the number Myra had written beside their address in Booysens, and got the disconnected signal.

He called Max Duma's office and was told Mr Duma was at a meeting, his secretary didn't know where.

It was too early for Joubert to have any information for him. There was no point in going to his chambers, since he'd cancelled all his appointments for the day.

He set out for home, but before he reached Parktown he changed his mind and headed south to Booysens. After consulting his road map and stopping at a café for directions, he arrived at Irvine Crescent.

It lay close to the railway line, a narrow street edged with neat bungalows. Neat, that is to say, except for Number seventeen which was an eyesore, the garden overgrown with weeds, the walls leprous with damp, several panes of the front windows broken.

Neil parked the car and forced open a gate that was rotting on its hinges. He climbed four steps to the front porch and knocked repeatedly on the door. Nothing happened. He started to walk round the building, and realized that he was being watched by a person peering over the fence of Number sixteen. The man was elderly, paunchy, dressed in green shorts and shirt, with a cotton hat pulled down over his eyes.

'No use makin' all that din,' the man said. 'You won't get an answer.'

Neil moved towards him, ploughing through knee-high weeds and nettles.

'I'm sorry if I disturbed you.' He reached out a hand. 'My name's Scallan.'

'Mercer.' A freckled paw brushed his fingers grudgingly. 'What d'you want?'

'I'm trying to trace a Mr and Mrs Alberton who lived here some years ago.'

The old man pursed up his mouth. 'Nobody of that name here.'

'Do you know who owns the property?'

'No, I do not. There's an old buzzard calls himself the care-taker. Caretaker my foot! I ask you, look at the place! Lets it go to rack an' ruin, an' all his muck blowin' over to my side. I've told the council the place is a health hazard, but they never do anythin'.'

'Can you tell me the caretaker's name?'

'Calls himself Smith. If you find him, you tell the bugger I'm after his blood.'

The old man stumped away, and Neil resumed his circuit of the house. A cat streaked past him, tail erect; a sleek animal, obviously well fed. He followed it, and rounded a corner in time to see it leap up a flight of steps to the back veranda.

Stooping to set down a plastic bowl was a scarecrow figure dressed in stained grey slacks and a torn jersey. Neil called out loudly, 'Mr Smith? May I speak to you, please?'

The man did not answer, but straightened up and began to move towards the back door in a rapid shuffle, shaking both hands above his head.

On impulse Neil called out again, 'Mr Alberton. I must speak to you.'

The man wheeled round.

'Go away! Get off my property!'

Neil advanced. 'My name is Neil Scallan, Mr Alberton. I'm David's father. I need to talk to you about your son.'

The haggard face contorted, the scarecrow arms lifted as if to call down vengeance from heaven.

'My son's dead. He's dead and gone. There's nothing to talk about.'

Neil climbed the steps. The cat, eating voraciously, watched him sidewise but did not leave its bowl. The man's arms fell, his mask of rage became a mask of grief. Neil put an arm round his shoulders and guided him into the house.

In contrast to the exterior, the interior was spotlessly maintained. A narrow passage led forward to what must be the living rooms, though the darkness of their doorways suggested that the drapes were drawn or the windows boarded up. To the right was a bedroom, and next to it a bathroom with an old-fashioned tub on curved legs.

To the left was the kitchen, where the old man evidently spent most of his time. It was large, dominated by an Esse stove that gave out a fierce heat. Polished pots hung from a tall stand, and food and cleaning materials were ranged on shelves covered with red and white oil-cloth. A refrigerator rumbled in a recess. In the centre of the floor was a deal table covered with newspaper, and on this were spread the parts of a model train – a Hornby, Neil thought.

'May I sit down?'

As Mr Alberton made no reply, Neil led him to a bentwood chair and sat down facing him.

'I didn't mean to shock you,' he said. 'Mr Mercer next door said the caretaker's name was Smith. I thought you were in Australia.'

Mr Alberton slowly raised his head. He seemed to have difficulty in focusing his eyes.

'I came back,' he said vaguely. 'A long time ago. Came home.'

'You went to Brisbane? Mark was with you?' Neil spoke quietly, hoping to coax some response from the old man.

'He was . . . at first. Then he found his own place, up the coast, at Keppel Bay.' Mr Alberton sighed. 'He liked the fishing. There was a company there, ran boats out to the Reef. Sold the catch. Mark got a job with them. He used to come to us, weekends. I had the service station going by then. Nothing big, but I was building it up. We were doing OK.' The old man's voice was suddenly defiant.

'I know what you're thinking; that Mark had plenty of money from the police, but that wasn't so. They paid his fees at the university for one year, and that's all. He didn't do it for the money. He hated the pinkos who were selling us out, he did what he could to stop them. He was a patriotic South African, but the Press hounded him. They wrecked his life. He had to leave varsity, there was nothing left for him here. We didn't have money, our name wasn't Scallan, we didn't have any protection. We had to get out of the country, Mark and Nesta and me. Mark never got over that. He never forgave what was done to him.'

Neil didn't argue the point. It was no use saying that Mark could have stashed money away in a Swiss bank or elsewhere. Mr Alberton had built up a comforting fiction about his son. It would be cruel to challenge it.

'Why did you come back to South Africa?' he asked.

This time the old man was silent for so long that it seemed he had retreated into a private world; but at last he whispered, 'Mark died.'

'I'm sorry. Very sorry. How did it happen?'

The opaque gaze veered away, searching some dark stretch of memory. 'There was an accident. He was run down. Killed.' Mr Alberton reached out to touch the parts of the engine on the table. Oil smeared his fingers and he wiped them carefully on the newspaper. 'You lost your boy,' he said. 'You know what it means.'

'Yes, I do.'

'After he went, there was no reason for me to stay. I wanted to come home, but Nesta . . . she wouldn't leave.

She liked it there. We split up. She married someone else, went to live in Perth. I came home. Been here ever since.'

Neil thought of the weed-grown garden, a barrier raised against intrusion, kindly or otherwise. It reminded him of Meg's flight from reality. Were the cases of David and Mark so different? Both had done what they believed to be right, both had paid an unbearably high price, both had left desolation in their parents' lives.

He said, 'Are you all right, living here alone?'

Mr Alberton looked at him. 'I'd be alone anywhere.'

There was nothing more to be said. Neil made his way back to the road. Reaching the car he saw that the legs of his trousers were covered with black-jack burrs, and pulled them off with savage jerks, as if they were legacies of the past.

13

The meeting of COVISA was set down for three o'clock. With an hour to spare, Neil stopped at a roadside café and bought a pie. Eating it in the car, he thought about the information he'd gleaned so far.

It wasn't much.

The photograph Anna had produced might or might not be significant. There was no knowing if Ernie Bryceland could enhance the print enough to allow identification of the man in the car; and if identification proved possible, there was no guarantee that it would throw any light on the Wivern note.

The interview with Ronald Alberton had been inconclusive. All he'd learned was that Mark Alberton was dead, and that if he'd been paid 17,000 rand, he hadn't told his folks about it.

Thinking of the money, Neil had to agree with Max Duma. It was an exorbitant amount to pay a university snitch. According to Myra, Mark had had little chance to spy on any major anti-government activist.

On the other hand, a number of academics had been murdered over the past two decades, and their killers had never been brought to book. One couldn't exclude the theory that the men on Wivern's list had been members of the hit squad.

Which brought one back to the key question: If Wivern had been involved in a group paid to carry out political assassinations, why in God's name would he reveal the fact now?

He . . . the person claiming to be Wivern . . . had said he was 'on his way out'. Out of the country? Out of this earthly life?

Neil rejected that idea out of hand. Wivern was not the sort to commit suicide.

But if he hadn't sent the note, who had?

The chief speaker at the afternoon's meeting was a visiting sociologist who talked about something he called the Continued Trauma Syndrome.

People, he said, who were subjected for long periods to the threat or the actuality of violence, were so severely traumatized that the damage became irreversible. Such victims – children who had seen their parents butchered, shack-dwellers who suffered repeated attacks by warlords or political firebrands – eventually lost the power to absorb punishment. They tried instead to expunge all memory of what they had suffered. They would not discuss it with one another, let alone with outsiders. Left in the danger-zones, they became unable to function in their day-to-day lives. Either they came to believe that their problems could only be solved by counter-violence, or they withdrew beyond anger to a zombie-like lethargy.

The solution, said the expert, did not lie in removing the causes of violence, or changing the constitution, or providing better housing and education. For the severely traumatized, there must be centres where they could be taught to rebuild themselves as human beings.

But what, thought Neil, if a whole nation has been traumatized, bludgeoned by threats year after year, its values distorted, its leaders discredited? Where does such a nation go for healing?

At the end of the meeting he tried to slip away quietly, but Gordon Leveret followed him into the foyer.

'Neil? Are you all right?'

'Yes, I'm fine.'

'You don't look it. Did you talk to the police?'

'To Joubert. He'll make some inquiries.'

'I see. Yes, I see. If there's anything I can do, Neil . . .'

'Thanks. I'll let you know.'

I am a victim. I cannot discuss what I feel with my own family, let alone with outsiders. Where can I go to be healed?

The thought of going home, of going through the charade of dinner with Meg, watching some mindless television show, was grotesque. He stopped at a callbox and left a message with Philomela that he'd be home late.

There was a light in Anna's apartment. He rode the lift up, rang the bell and stood leaning against the wall. She came to the door with a book in her hand. Behind her a compact disc was playing *Der Rosenkavalier*. Neil stepped forward and put his arms round her. She lifted her face but he did not kiss her. The warmth and familiar scent of her body roused no response in him. He was in the grip of a lethal inertia.

'You're cold,' Anna murmured. 'You're shivering. Come in.'

He let her lead him to the sofa. She sat beside him, chafing his hands. 'What is it, are you ill?'

'Point of no return,' he said. His voice sounded distant in his own ears. 'I can't go back, and if I go on, you could all be in danger. You. Meg. Todd.'

'What danger? What are you afraid of?'

He shook his head, unable to put his fear into words. 'I can't talk at home,' he said. 'Meg mustn't know, and Todd evades the issue.'

'You can talk to me.'

'Can I? Yesterday you seemed . . . angry. I thought you didn't want to be part of it.'

91

'I was angry for you, not with you.' She thought for a moment, reaching for the right words, then said simply, 'I love you, Neil; I can't help being part of what you are and what you do.' Freeing her hands, she went to the sideboard and poured brandy into two glasses, handed one to him.

'Drink up.' As he obeyed, she sat down again. 'Tell me what you're afraid of.'

The brandy was fire in his throat. Closing his eyes he said slowly, 'Something that existed before David died, that killed him, that still exists. Something that has eaten away the structures of society and of justice, so that those we appoint to protect us can no longer do so.'

'An organization? What sort of organization?'

'I don't know.' He opened his eyes. 'I saw Mark Alberton's father this morning.'

She blinked. 'I thought they emigrated to Australia.'

'They did. They went to Brisbane. Then Mark was killed in a car accident, the parents divorced, and the old man came back here. He's living at the house they owned before. We talked.'

He repeated his conversation with Mr Alberton, and as he did so the floodgates of memory burst. He spoke of David, recalling long-forgotten incidents, things that David had said and done and thought.

Anna brought him food and he ate without tasting it, rose from his place and paced about the room, talking, talking. At last he reached a standstill in mind and body.

'What time is it?' he said.

'Almost midnight.'

'God, I must go!'

She walked to the door with him. He stood looking down at her, seeing her not with the eyes of long habit, but as others must see her: middle-aged, her skin still rosy but marked by small lines, her hair greying at the forehead line, her eyes fixed on him with a kind of steady compassion. He had asked too much of her, all these years. He had made her vulnerable, protected everyone but her.

He sought for words, and at last said quietly, 'I can't survive without you, Anna. You know that?'

She nodded, half-smiling, pushed him through the door and gently closed it.

His was the only car left in the visitors' parking area. In the past he'd been discreet about leaving it there. Now he didn't care who saw it.

The catharsis of talking to Anna had restored his clarity of mind. It had brought David back. Not the David who'd been a victim, but the David who'd been assertive, shrewd, perceptive in his judgement of other people.

Dave had been a hard man to fool.

He would not have been fooled by Mark Alberton. He would have picked up Mark's deception. He would have watched him, knowing him to be an informer.

Perhaps that was the reason Dave had been murdered.

On arriving at his chambers the next morning, Neil found a large buff envelope on his desk. It was from Ernie Bryceland and contained a dozen photographs, blowups of the snapshot Anna had given him.

He carried the photos to the window and examined them carefully. Most of them he discarded at once. The process of enlargement had improved the image of Silbermann, but turned the figure in the parked car to a grey blur.

Two of the prints were enhancements designed to brighten the picture, and these he studied closely. They showed a man leaning towards the car window. The face was in three-quarter position, the features small and puglike, the lower jaw jutting, the nose short and blunt. The man wore an overcoat – it had been a cold winter, Neil remembered – and a black felt hat pulled well down on his forehead. Despite the smudginess of the print, the impression conveyed was one of power and tenacity.

Neil returned the pictures to the envelope and locked that in his briefcase. He spent the rest of the morning working on an opinion, and at 12.30 walked the short distance to the headquarters of the ANC. Max Duma was in a meeting, but emerged at once in answer to Neil's message.

'Neil, hi man! I was going to phone you this afternoon. Come to my office and I'll tell you.'

When they were alone, Duma lit one of the Gauloises cigarettes he favoured, and waved a paw through the smoke.

'I asked around,' he said, 'spoke to some of our people who were here when David was killed. They knew all the men on your list. They have a file this thick on Hermanus Silbermann. He was your old-fashioned pig, know what I mean? Used to boast about how many kaffirs he'd donnered. He was the subject of an internal inquiry, once, but he was lucky. He stayed on the Force, even got promoted.

'He was a good boxer, crack shot, won cups for it. Our people reckon he was with the dirty tricks department for quite a few years. During 1983, the Press and the Opposition got him in their sights. He didn't enjoy that. He had high blood pressure and in January 1984 he had a stroke and had to leave the service.

'Next on the list, Sergeant Lemuel Moroka, alias Landros Sibisi. A Tswana, joined the police when he was eighteen, and was on the take from the start. If you wanted your shebeen to stay open, you had to be nice to Moroka. If you needed a heavy-duty licence or a trading-permit, Moroka could tell you who to pay, and of course he took his cut.

'He was near to being kicked out, but someone thought he was too good a skelm to waste. He was sent on a special combat course and spent time on the border. When he came back he was posted to Silbermann's team. In January 1984 some people called at his house in Sophiatown and chopped him down to size. Nobody cried for him.'

Duma took a last drag on his cigarette and stubbed it out. 'We don't have much on Wivern,' he said, looking sideways at Neil. 'Everyone says he's dead.'

Neil took Bryceland's photographs from his briefcase and handed one to Duma.

'The man in the car? Do you know him?'

'Not offhand.' Duma was frowning. 'Where was this taken?'

'Outside the court where they held the inquest on David. I thought it might be Silbermann's controller.'

'No. Those hyenas avoid the light of day.'

'Who, then?'

'SB, maybe. A technical expert, maybe. A lawyer, maybe. Perhaps Silbermann needed to be coached before he appeared in court. Can I borrow this?'

'Sure.'

Duma put the photograph in his pocket. 'I'll let you know if I learn anything, but it's a poor picture so don't hope for too much.'

14

'I've nothing to report, I'm afraid, Mr Scallan.'

Major Joubert sounded genuinely apologetic. 'I had the note checked for fingerprints. One set showed . . . yours, probably . . . certainly not Wivern's. I spoke to a senior officer at headquarters. He told me categorically that Wivern is dead and therefore the note is a fake.'

'Did you tell him I want the faker identified?'

'Yes. They'll do their best, but they doubt if they'll have any success. There's very little to go on. They feel . . .' Joubert broke off uncomfortably, fingering his glasses.

'Yes?'

'I'm sorry, sir, but they feel you're . . . not able to be impartial about Wivern.'

'They think I'm obsessed.' As Joubert made a deprecatory gesture, Neil insisted, 'They think I'm obsessed. Do you, Major?'

Joubert coloured. 'No, I don't. Whether Wivern's alive or dead, someone is using him to harass you and disturb your work with COVISA. To me, that merits an investigation. I took it on myself to make a few inquiries,' he opened a drawer and drew out a sheet of paper, 'about Wivern's associates. According to our records, Wivern died in February 1984. Lemuel Moroka was murdered in January 1984. Colonel Hermanus Silbermann suffered a massive stroke in January 1984 and was invalided out of the Force. And Mark

Alberton left the country in September 1983. Of the four members of Wivern's group, one is retired and two are dead.'

'Three,' Neil said. As Joubert looked startled, he repeated, 'Three are dead. Mark Alberton was killed in an accident in Australia.'

'How do you know?'

The form of the question was odd. Joubert evidently knew of Mark's death.

Neil said nonchalantly, 'An acquaintance told me he'd died.' He carried the contest to Joubert's ground. 'How much was he paid for his services to the police?'

Joubert didn't like that, but he answered without hesitation, 'His education was covered for a year.'

'Then where did he get the seventeen thousand rand?'

'If he did get it, it wasn't from police funds.'

Neil held Joubert's gaze. 'Let's assume the note is accurate. It sets down precise amounts of money. Would the members of a hit squad know what the other members received?'

'I've no knowledge of hit squads, Mr Scallan.'

'Of course not. Still, I suppose it's possible such men might discuss fees among themselves? Silbermann was in command of the team. He could have known the figures. So could Wivern. There's a third possibility. The note was written by someone who worked for the same covert operation – a controller, perhaps, or a paymaster.' Neil opened his briefcase and took out Bryceland's photographs. 'Take a look at these, Major. Tell me if you recognize the man in the car.'

Joubert studied the pictures. 'Where did you get the original?' he asked.

'From an old file at the university. It was taken at the time of the inquest on my son. Can you identify the man?'

'No, it's no one I know.'

'Will you make inquiries?'

'Certainly.' Joubert took one of the prints.

'One thing more,' Neil said. 'Does the name MacAllivan mean anything to you?'

'Spell it, please.'

Neil did so, and Joubert went to the computer keyboard, tapped keys, waited for the display on the screen. He drew a blank, tried alternative spellings, and turned back to Neil.

'Can you tell me in what context the word was used?'

'Guns. Someone was talking about guns and used the name.'

Joubert's brow cleared. 'Ah, that's different.'

He pressed more keys. This time a line drawing appeared on the screen.

'There's your Mac Eleven,' he said. 'It's an American rapid-fire pistol, precision made, light enough to pack in that briefcase of yours. It's used by the US anti-terrorist teams.'

'And in this country?'

'No way. The arms embargo prevents that.' Joubert came back to his desk. 'I'd like to know your informant's name, sir.'

'I'm afraid I can't recall. It was just a passing remark, at a dinner party. Major, I'd like to pay a visit to Colonel Silbermann. Is that possible?'

'He's in a private nursing home for the frail aged.' Joubert thought a moment, then scribbled on a slip of paper, pushed it over to Neil. 'I don't know if he's well enough to receive visitors. You'd better phone first.'

'I will. Thanks, Major, you've been very helpful.'

Neil left. Joubert reached out to press a buzzer under his desk.

The man who entered the office was small and dapper, dressed in a dark grey suit, a white shirt with a fine grey stripe and a grey silk tie. His pepper-and-salt hair was longish, combed straight back. There was a wart at the right side of his mouth. His eyes, behind heavy-framed spectacles, had

98

the cold alertness of one who is used to being lied to. Selwyn Marx looked like a senior investigator from the Department of Inland Revenue. He had in fact been high up in the Fraud Squad before he was put in charge of the CID's new technology, with the rank of colonel.

Joubert said anxiously, 'You heard all that, sir?'

Marx nodded. 'Mr Scallan isn't telling us all he knows.'

'I should have pressed him to name his sources.'

'No. He wouldn't have told you anything, and he wouldn't have come back.' Marx reached out a hand for the photograph Neil had left, and stared at it with mute concentration.

'The man on the steps is Hermanus Silbermann.' Joubert found his superior's silences unnerving. 'Scallan wants us to identify the man in the car. Do you know him, sir?'

Marx shook his head. He tapped the print. 'There's a patrol car blocking the far end of the street, see? The courthouse must have been cordoned off. This car's parked on the yellow line right in front of it. Must have been someone with clout.' He paused. 'He seems very interested in Silbermann.'

'Scallan suggests the man could be a controller. He's still talking about hit squads. I told him . . .' Joubert met the colonel's limpid gaze, and finished lamely, 'I told him I know nothing about any of that.'

'Ignorance is bliss.' Marx laid the picture aside. 'But what's your true opinion, Tiaan? Did such squads exist? Hmn?'

'There were . . . unexplained deaths. Unsolved murders.'

'What the Americans euphemistically term "executive action".'

'Yes, but that doesn't mean there was a deliberate policy . . . after all, there are rotten apples in every barrel.'

'Even ours. Were Silbermann and his crew such apples? Did they profit by their rottenness? And if so, who paid them? Those are the question Mr Scallan poses. We have to find answers, if we can. By what you say, headquarters was unhelpful.'

'Very.' Joubert looked unhappy. 'In fact, they told me to get lost. I'd have thought someone would want to interview Scallan. The last thing we need is for him to cry "cover-up".'

Marx examined his fingertips. 'Perhaps you spoke to the wrong people. Things are not straightforward, these days. Not everyone on the Force is a starry-eyed reformist. Many cling to the old ways. A few have skeletons they'd like to keep hidden.'

'There's an amnesty. They can apply for indemnity.'

'Ah, that amnesty! A law that allows crooks to confess in private, so no one will ever know for what they're being pardoned. The trouble is, that law won't be allowed to stand. It will be repealed as soon as a new government takes office. The guilty know that, so they won't confess. Not yet. "Oh God, make me good, but not yet!"'

'You take Scallan's story seriously, sir?'

'I take everything seriously, Tiaan. I have that sort of mind. Neil Scallan is an important man, a member of an important committee. If he's been harassed, that's already a serious matter; but if it turns out there's truth in that note, if Wivern is alive, then we could be looking at a highly dangerous situation.'

'So what do you want me to do?'

'Leave things to me, for a while. Play along with Scallan, be helpful to him. Can your computer boil a kettle?'

As Joubert stared, Marx gave his frosty smile. 'I'd like a cup of coffee. We can run the tape again while we wait.'

Joubert ordered coffee and started the tape he'd made of his conversation with Neil Scallan. Marx listened, hands folded under his chin. At the end he sighed.

'Guns and money. The twin sources of power in a violent world.' He became brisk. 'I want to know everything there is to know about Wivern, in life and in death. I want a follow-up on that mention of a Mac Eleven. I want to identify the man in the photograph. Most of all, I want to know who's trying to pull our strings. I don't like being manipulated. I'll ask questions. You play it by the book, hear me? Keep in touch with Scallan, and keep me fully informed.'

A tray of coffee was brought. Handing over a cup, Joubert said, 'Scallan plans to visit Silbermann.'

'Let him, it can do us no harm.' Marx shovelled three spoonfuls of sugar into his coffee. A sweet tooth was his single self-indulgence.

15

The Orchardside Residential Home for Retired Citizens smelled of new-cut grass, pine needles and money. Round a central one-storey block were clustered some twenty cottages, each with its own pocket-handkerchief lawn and neat flowerbeds. To the right of the driveway several old codgers were playing bowls on an immaculate green, and to the left, in a parking area ringed by indigenous shrubs, stood a BMW and a Mercedes.

It looked, Neil thought, way beyond the pocket of a pensioned-off colonel of police.

The prospect of meeting Silbermann again made his palms sweat. The man had typified the old-style security policeman: brutish, sure of his right to threaten, imprison, even torture those he saw as enemies of the State.

Perhaps he felt secure now, behind the protective arm of the Further Indemnity Act of 1992.

It would be a pleasure to disillusion him, to inform him that without doubt the Act would be repealed; that before he could hope for indemnity he must be judged by a tribunal and his offences made known to the nation. Let Silbermann feel some of the fear he had inflicted on so many others.

The entrance hall was pink-walled, furnished with comfortable chairs and a natural-oak coffee table. At the reception desk a receptionist in a fuchsia uniform was reading *Lace*. Neil cleared his throat and she glanced up.

102

'I've come to visit Colonel Hermanus Silbermann,' Neil said. 'My name is Neil Scallan.'

'Are you a relative, Mr Scallan?'

'No. We were close in the past. I happened to be in the neighbourhood and thought I'd look him up.'

She nibbled on a nail, then shrugged. 'I suppose it can't do any harm.' She pressed a button on the counter. A nurse-aide in a lavender overall appeared from a side office.

'Is Colonel Silbermann awake?' the receptionist asked.

The girl nodded. 'He's out in his chair.'

'Take this gentleman to him, please.' The receptionist returned to her paperback, and Neil set out after the nurse. They walked along a corridor and pushed through a swing-door into a hospital ward. The acrid smell of medicines and ammonia told Neil that this was the frail-aged section of the Home, and he said quickly, 'Is Colonel Silbermann well enough to have visitors?'

For answer the nurse opened a door and led him into a small sunny patio. In its corners stood Ali Baba pots filled with pelargoniums. Along the far wall was a line of wheel-chairs, in which sat men so old, so unmoving, that Neil wondered if they were still breathing.

He scanned the faces, the greyish flaccid skin and leaden eyes, and decided there had been a mistake. Silbermann wasn't here. Yet the nurse was heading for the farthermost chair, waiting beside it with an air of patient forbearance. Neil went to join her.

Silbermann had been a huge man, six foot four with the build of a lock-forward. Now he was skin and bone, the face a mask, cheek- and eye-muscles dragged downwards, drooping lips cyanosed.

The nurse put her mouth close to the old man's ear.

'Colonel? A friend to see you, dear.' As there was no response she put a hand under the sagging chin and lifted.

Neil gazed into eyes that held no sign of recognition, or indeed of any intelligence.

'Since his second stroke,' the nurse said, 'he doesn't know people any more.'

Neil bent forward. 'Colonel Silbermann, it's Neil Scallan. Do you remember Neil Scallan?'

The eyes remained blank. A line of dribble rolled from the corner of the mouth. The nurse stretched out her hand again but Neil said sharply, 'No, leave him be.'

They made their way back to the foyer. The receptionist gave Neil a bright smile.

'Did you see your friend?'

'Yes, thank you. I'm afraid he didn't know me. Is there any member of his family I could talk to?'

She shook her head. 'His wife's dead, and they never had any kids.'

He wanted to ask who paid Silbermann's bills, but knew that that kind of information would not be forthcoming. Doubtless some firm of attorneys looked after the old man's affairs.

He returned to his car and stood beside it in the sun, reviewing his options. They were few. If Silbermann had ever worked with a hit squad, he would have kept the fact to himself. Any blood money he'd received would have been carefully laundered and salted away in some innocuous bank or pension fund. As for his mind, that was blown. Nothing could be learned from him.

Four men had been implicated in Dave's death. Of the four, three were dead, and one had vanished into hopeless senility.

In a two-star hotel bedroom on the west side of the city, a telephone rang. The man lying on the bed rolled over and lifted the receiver.

'Luther.'

A wheezing voice spoke: 'You're going to Muller's Drift in Natal. You know it?'

Luther did not answer, and the voice said impatiently, 'Did you hear what I said?'

'I heard. What's the job?'

'You go to the hotel, sign in as Leonard Brown, ex-army, unemployed. Make yourself known to a botanist staying there. He works for the Harleigh Institute as a field researcher.'

'What's his name?'

'Todd Scallan. He's Neil Scallan's son.'

Luther swung his feet to the ground. 'No way. I don't take out any Scallan; it's too dangerous.'

'Who said anything about taking him out? On the contrary, your job is to keep him in, keep him alive! For double your last fee.'

Luther was scowling. 'If this is about Wivern . . .'

'It's about you, dear boy.' The voice became wheedling. 'Your chance to finish that business, once for all. Think of the money, Luther. Think of South America.'

Another silence. Then, grudgingly, 'OK, tell me what you want me to do.'

The voice told him.

Neil arrived home soon after five o'clock. As he came through the front door, Meg hurried down the stairs, halting on the lowest step. She wore a so-far-and-no-further expression.

'Someone phoned about half an hour ago. He wouldn't leave his name, he said he'd call again later.'

'Good. Thank you.'

'It sounded like Louis Baird's voice. I identified myself, but he rang off. I think it's the height of rudeness when someone I know won't give his name.'

'Louis's in the middle of an Arabian oilfield. Perhaps he was cut off.'

Neil started up the stairs, wanting a bath and shave, wanting time alone. Meg followed him, still spoiling for a fight.

'I dislike Louis Baird. I think he's a crook.'

'No. He comes close at times, but he stays within the law.'

'You're so naive about your friends. I wouldn't be surprised if he deals in drugs.'

'He wouldn't dream of such a thing.'

'How do you know? How do you imagine he's made all his money? A little snot from Roodepoort? No one knows what he really does for a living.'

'He watches money, his own and other people's. He's a financial genius.' Neil reached his bedroom and began to strip off his clothes, hoping that that would drive Meg away. She disliked being reminded of their former intimacy. Tonight, though, she stayed.

'What's he after?' she demanded. 'I suppose he wants you to do some of his dirty work!'

Neil bit back an angry reply. 'He's returning my call. I need to talk to him.'

'What about?' Meg edged round so that she was facing him.

'A business matter. Money.'

'Don't try to fob me off! I know you're up to something, I can always tell. What money are you talking about? Is it for that woman?'

Looking at her, Neil saw that her skin was slick with sweat and she was trembling. In a moment she would become hysterical. He forced himself to speak quietly.

'There's nothing to worry about, Meg. I want to ask Louis about certain financial transactions for a client. I need his advice, that's all.'

She remained staring at him for a long moment, then made

for the door. Reaching it, she turned back. 'I'll be out for dinner. Aline and I are going to a concert. I'll have something to eat when I get back.'

She went away and a few minutes later he heard the front door slam.

He took a shower and was dressing when the bedside telephone rang.

'Neil? It's Louis. Francesca said you wished to talk to me.'

'Where are you speaking from?'

'Johannesburg. I'm at the Carlton.' There was amusement in Baird's lazy voice. He enjoyed catching people off balance. It had helped to make him one of the richest men on the planet.

'I need your help,' Neil said. 'An urgent personal matter.'

'Ah. Come to dinner, then, and we'll discuss it. Eight o'clock suit you? Good. Arrivederci, my friend.'

16

In Muller's Drift the heatwave intensified. The veld was leprous grey, the lichen rusty on the rocks. Even the moon seemed distorted, rising swollen-cheeked above the plain. Thulani eyed it uneasily.

'Perhaps it will storm. We should go by daylight.'

'It will not storm tonight, and some muti must be gathered by the light of the moon.'

Thulani stowed a torch in the canvas pouch at his waist. He had told his grandparents he would be helping his uncle to label medicines, and would spend the night at Siphiwe's house.

When the moon was fully up they set out, circling round the village, following the hill paths. Siphiwe leaned on a tall staff, Thulani carried two fighting sticks. Sometimes they passed other kraals and the thin dogs barked. Below them, the lights in the village went out one by one, leaving only those at the railway and police station.

Siphiwe stopped often to pick herbs. 'When the big rain comes all these will be washed away, as they were in the last great storm.'

'The radio didn't say there would be another like that,' Thulani said. Siphiwe regarded him with scorn.

'What does that little box know? People draw lines on a piece of paper and say this will happen, that will happen. It never happens. I don't draw lines. I can smell rain coming,

108

I can read the signs in the sky. I can throw the bones. When it comes the storm will be very big, it will break bridges and houses and drown men and beasts. That will happen.' He collected a strip of bark from a tree and tucked it into the bag he carried. They worked their way slowly along the ridge.

'That's the place,' Thulani said at last. 'We'll have to climb down. It's very steep.'

'There's an easy way,' Siphiwe answered. 'I'm an old man, I remember when this land was ours. I walked here when I was your age. We go this way.'

He started down a slope that grew rapidly steeper. Thulani followed, feeling the parched grass stab at his ankles. They reached a barbed-wire fence and climbed through it. Far away to their left, a dog barked. Closer they heard a pump engine thumping rhythmically. They struck a path that presently joined a cart track. Thulani hung back, nervous.

'This place is tagati, Uncle. Forbidden.'

'Is that what they teach you at your big school?' Siphiwe laughed, shaking his head. 'Ai, no, boy. They say it's tagati because they want to keep us from what is ours. Come.' He pressed forward, swinging his blanket high round his shoulders so that he resembled some humped animal. Thulani followed him along the track. He could hear the pump more clearly, and the burble of the stream that fed it.

Siphiwe held up a hand. 'There'll be a man watching the pumphouse,' he said. 'We must go quietly.' He left the track and climbed a knoll, crossed the stream and entered the forest. A path led them upwards. The trees grew thickly, and sword fern and creeper encroached on the trail, sometimes forcing them to bend almost double. The foliage blocked the moonlight and Thulani took out his torch to guide them.

At last they reached a clearing, a natural arena in the gorge. To their left, the rock face was tree-grown, vegetation clinging wherever there was enough soil. To their right, the stream fell into a wide pool, almost empty now. Beyond it was a slope of grass and the mouth of what looked like a large cave.

There was a stillness about the place, a thick silence that raised the hairs on Thulani's neck. He clutched at Siphiwe's arm.

'Uncle, there's someone here. We're being watched.'

'The spirits of our ancestors are close. They won't hurt us.' Siphiwe pointed ahead. 'Is that what you're looking for?'

Thulani shone the torch on the wall of greenery that barred the path. Five metres tall, it spread an umbrella of fronds against the sky. He stumbled towards it, his feet slipping on the masty ground. Reaching out, he felt the prickly tip of the lowest frond.

'This could be the right one,' he said. His voice was shaking with excitement, and Siphiwe said disparagingly, 'You can't make muti from it. It's useless.'

'I must cut leaves,' Thulani said. He bent to lay his sticks on the ground and saw a scattering of seeds at his feet. Picking up a handful he thrust them into the pouch at his belt. He was reaching for his knife, to cut a leaf, when he heard Siphiwe call out sharply, 'Hai! Who are you? What do you want?'

The men rushed at them from the shadows, brandishing cane-knives. They cut Siphiwe down, and ran at Thulani. He snatched up his sticks and tried to defend himself, but he was driven back to the rock face. A panga severed his hand from his wrist, another sliced at his throat. He fell before he could utter a cry.

The white man moved to the nearer of the two bundles lying in the long grass, and twitched back the sacking that covered

110

it. He studied the hacked face with distaste and turned to the black man standing a pace or two away. He said in Zulu, 'Do you know who this is?'

'Yes. He is Siphiwe Msomi, the herbalist.'

'And the other?'

'The grandson of his brother Hosea Msomi.'

The white man bent to fumble through the pouch that lay next to the dead man.

'What's all this stuff?'

'Medicines. They were gathering medicines. They were doing no harm. You should have left them alone. It's bad to kill a witch doctor.'

The white man stepped back, wiping his hand on a tuft of grass. 'I know what my orders are.'

'There'll be trouble. The police will make big trouble.'

'Over a kaffir witch doctor? Don't talk crap.'

'The boy worked for the man from Igoli – Mr Scallan.'

The words were spoken with malicious edge, and the white man swore, moving over to examine the second body. After a moment he straightened.

'Tell your boys to put them in the truck. We have to move them away from here.'

'Why, what difference does it make? They won't talk.'

'Do as I say, hear me? Put them in the truck and cover them over. You ride in the back with them. If anyone stops us and asks questions, I'm taking you to the doctor. You're very sick in your belly.'

'A sick man should ride in the cab with the driver.'

'Move, fast, or I'll give you sick!'

The black turned away and raised his hand. Figures emerged from the shadow of the trees, lifted the bodies into the truck and covered them with sacking. That done, they withdrew, huddling together, fearful of the baleful moon and the dark forest behind them.

The black man vaulted to sit on the metal bench in the back of the truck.

'Where will you take them?'

'The Smedleys' place.'

'That's foolish. The police are watching there.'

'Well then, where must I go?'

'Put them on Khumalo land. The police will think it was a fight between the Msomi and the Khumalo people. They're always fighting.'

The white man turned to stare at the labourers near the trees.

'Will those dogs bark?'

'No.' The black man laughed. 'They'll run away. By tomorrow night they'll be back in Transkei or Venda, shaking in their rat holes. I will neither talk, nor run. Now you'd better hurry. Someone may see us here. The big boss wouldn't like that.'

'He's not here.' The white man's face contracted in terror, and the black man eyed him with contempt.

'He'll come soon. Before the storm.'

The white man climbed into the cab and started the engine. The truck moved slowly along the rough track. It followed a circuitous route across the plain to the land east of the Msomi territory. At times the uneven ground made the bodies on the floor stir, and the black man put out a foot to steady them.

17

'If Wivern's alive, which I doubt,' said Louis Baird, 'he's a very small needle in a very big haystack. I don't think I can find him for you.'

Seated in an armchair, facing a picture window that over-looked the constellar lights and ant-world traffic of central Johannesburg, Baird was far from being the little snot of Meg's description. His authority was quite disproportionate to his small size. His wedge-shaped face with its luminous eyes and pointed chin might be gnome-like, but the gnome hailed from Zurich, not Roodepoort.

He'd been eight years old when Neil first met him, an oddity plucked from his primary school and dumped down among boys twice his age because of his formidable intelligence. Neil had been detailed by the headmaster to see the newcomer wasn't bullied. It soon became obvious that little Baird could look after himself. He was tough, shrewd, had a street-fighter's vocabulary and a gift for clowning that his schoolmates admired. He became their mascot, famed for his luck. Time showed that the luck could be attributed to a genius for gauging the odds.

When he was fourteen, Louis' father was killed in a mining accident, and his English mother took him back to London. Six years later he appeared in Neil's chambers, already rich and looking for ways to be richer. 'I want an adviser,' he told Neil. 'Someone I can trust absolutely.'

'I know nothing about finance, Louis.'

'Don't worry, I'll look after the money. Your job is to keep me out of gaol.'

There was no formal agreement. Baird consulted Neil about his mushrooming interests in Southern Africa. If a deal looked like coming too close to piracy, Neil issued a warning and Baird made corrections. Sometimes he suggested an investment. His tips were always rock solid.

Within ten years Baird was a multi-millionaire, and within twenty he was a financier of international repute. Inevitably his success took him into the realms of the mega-rich and powerful; yet he continued to visit South Africa, and never failed to arrange a meeting with Neil. Neil believed their unlikely friendship endured because Louis needed a confidant who had no axe to grind.

Having listened without interruption to Neil's account of the events of the past few days, he set aside his empty brandy glass and said gravely, 'How sure are you that it was Wivern who made that call?'

'Not sure at all. At first, I was convinced it was his voice, but now I've had time to reflect. The accent was the same as his – the Malmesbury burr – but the tone quality was wrong.' He met Louis' thoughtful gaze. 'I hated that man, you understand? I hated his guts. I haven't forgotten anything about him. I don't think it was Wivern who spoke to me.'

'You think he arranged for someone else to make the call? Why would he do that? Why would he wake the sleeping dog?'

'That's the crunch question. Perhaps he plans to apply for indemnity. He knows I'm on COVISA, dealing with the amnesty law. Perhaps he fed me information, hoping that would help his case.'

'Possible, I suppose.' Baird ran a finger over his chin. 'But if Wivern's dead, there has to be another explanation. We must accept that someone posed as Wivern, made the call and left the note suggesting that David was murdered by a hit squad. One can propound reasons. Perhaps the caller holds a grudge against one or all of the men on the list . . . wants their crimes exposed, but isn't in a position to do the exposing himself?'

'What would be the point? Alberton and Moroka are dead, and Silbermann's a basket case.'

'People aren't above harrying the dead. The Jews are still hunting down men whose crimes are fifty years old. But I agree, this thing hasn't the smell of righteous indignation. Do you think someone hopes to make political capital out of the exposure? The ANC and the far right would relish proof that the government or its agencies sanctioned hit squads.'

'That's possible, but I'd have thought they'd go about it in a more direct way.'

'We do seem to be dealing with a tortuous mind.' Louis was watching Neil closely. 'Perhaps the caller expected you to raise a public outcry. You didn't, Neil. Why was that?'

'I couldn't embroil my family.'

'Understandable, of course. How long have you believed that Wivern was responsible for David's murder?'

The question surprised Neil, but he answered it honestly.

'Almost from the beginning. For a few weeks after he died I couldn't . . . think straight. I took it the police inquiry was genuine. As time went on, I realized that Silbermann was a thug, and Wivern a sadistic crook. I felt in my gut that he set that bomb. I did my best to prove it, but I couldn't find any evidence to support me. Even my friends thought I was deranged. Last year, when the truth about covert operations began to spill out . . . when MI was investigated, and the CCB . . . my view was to a large degree confirmed.'

'In other words, the contents of the note are not as

incredible as they would have been a few years ago. Yet the police still think it's a hoax? Of no importance?'

'So I'm told.'

'Strange.' Baird's luminous gaze shifted to the window, as if he contemplated some distant point in the sky. After a while he sighed.

'You mentioned a Mac Eleven,' he said.

'Yes. It's probably irrelevant . . .'

'Guns are never irrelevant.' Louis picked up the brandy bottle, poured a measure into Neil's glass and his own. 'In my work,' he said, 'guns are very much to the point. The violence here, for example, is costing me a mint of money, because my concerns need a stable economy in which to make profits. That's true for most financiers. But there are exceptions. The drug-pushers, the arms-dealers, profit from a climate of violence. They need it. They promote it. Are your learned colleagues on COVISA aware that the supply of arms to this part of the world has now entered phase three?'

'We know there's been a considerable escalation in the number of guns in circulation.'

Baird made an impatient movement. 'It's not a question of quantity only, but of quality. The type of weapon involved. The people distributing, controlling, using weapons. Let me explain.

'Phase one covered the period when the purchase, possession and control of weapons was confined to the government and its security forces. Farmers and a few private citizens owned guns, criminals used them to a limited extent; but the victims in the casualty wards carried injuries made by knives, or sticks, not guns.' Louis' smile was lopsided. 'Those were the good old days.

'Phase two began when the world imposed sanctions on the sale of arms to South Africa. The South African government responded by developing its own arms factories, and by procuring supplies from less scrupulous sellers. At the same time, the liberation struggle within this country intensified, broadening the local market for guns, grenades and explosives. As the wars of resistance to our north ended, discharged soldiers in need of cash sold their surplus arms to whoever was ready to pay. Guns began to feature regularly in bank raids, township battles, attacks on political rivals. That was the age of the AK47.

'Phase three, ironically enough, arrived with the official end of apartheid. The power struggle between political factions created a lust for weapons. The armed wings of political parties and of black and white extremist movements began to build up caches of arms, and to train their supporters in their use.

'It was at this stage that reports began to come in from monitors here and abroad that the weapons crossing the border were no longer the debris of old wars . . . much of which was obsolete or ruined by neglect. The equipment was modern sophisticated stuff . . . such as, for example, the Mac Eleven rapid-fire machine-pistol mentioned by your friend Sergeant Ritter. Phase three, in short, is the phase of low-key civil war.'

'You're very well-informed, Louis.'

'Yes. In my own interests, I have to be.'

'Do you know who the suppliers are?'

'There are dealers in North Africa and the Middle East, also in the dragon states . . . Seoul, Hong Kong, Singapore, Bangkok. The trouble is that as soon as we locate and destroy one cartel, another springs up to replace it.'

'Why can't the trade be stopped at source, with the arms manufacturers?'

Baird gave a dry laugh. 'Because they're too big, too influential. They control governments, not the other way about.

We do our best. It's not good enough.' He smiled at Neil. 'To return to George Wivern . . . I doubt if I can help you, but I'll pass the word. If he's alive, and has a criminal record – particularly if he's into drug-trafficking or terrorism – he may be on file in Washington, London or Wiesbaden. I'll ask around. If anything comes to light, I'll let you know at once.'

It was after midnight when Neil left Baird's suite. He felt bone-weary and depressed, no longer certain whom he could trust. Even Louis' arrival in Johannesburg troubled him. It seemed too fortuitous. Had he come out of friendship, or concern for his shadowy business affairs?

Everyone seemed to be working to a hidden agenda. The spectres of the past combined with the faceless menace of the present to oppress him.

He drove too fast through the emptying streets. The house was in darkness save for the security lights at front and back. Letting himself in through the garage door, he climbed the backstairs and made his way to bed.

The alcohol he'd consumed did nothing to help him sleep. He tossed and turned till the small hours, then dreamed of a dragon with the smiling face of Wivern, which wrapped its coils about him in an unbreakable stranglehold.

18

As Todd left the dining room next morning, he encountered Sergeant Ritter.

'Mr Scallan, can I have a word with you, please?'

'Sure. Would you like some coffee?'

'No, sir, thank you. It's about the boy who works for you. Thulani Msomi.'

'What about him?'

'He's dead, Mr Scallan. A herd boy found him and his uncle in a gully, early this morning. Stabbed to death, both of them, throats cut.'

As Todd stared blankly, Ritter took off his cap and wiped his forehead. 'I wonder would you come over to the station with me, meneer? There's a couple a things I'd like to clear up.'

'Where . . . where did this happen?'

'Over on the east side, 'bout five kays from here.'

'But that's Khumalo land. Siphiwe wouldn't have gone there, specially at night. He wouldn't have taken the boy . . .' Todd couldn't focus his thoughts. He said uncertainly, 'I should go and talk to Thulani's family.'

'They already at the station, Mr Scallan. They came to identify the bodies. You can talk to them there.'

It was an order, not a request. Todd walked with Ritter to the police station. A police van was drawn up at the door, and a black constable was talking to a small group of labourers. Ritter led the way into the building. The charge office was crowded with people, most of them black. A tall headman was arguing fiercely with a desk policeman, making wide gestures of denial. Next to him was a child of about seven, barefooted, tearful and obviously terrified. Todd guessed he was the herd boy who'd found the bodies.

Ellen and Hosea Msomi were in Sergeant Ritter's office, the old man sitting bolt upright, his near-blind eyes staring straight ahead. Ellen turned her head as Todd came towards her. Tears rolled down her face, she brushed them away continually with a folded handkerchief. Todd bent and put his arm round her. She rocked from side to side, moaning softly.

Hosea struck the desk with the flat of his hand.

'It is lies,' he said fiercely. 'They would not go on to Khumalo land. Khumalo and I have an agreement. Siphiwe knew of it, he would not have broken it. He never broke our laws. You know that.' He turned from Ritter to Todd. 'You know that, Inkosi. Siphiwe would have stayed on our land.'

'Yes.' Todd took the chair Ritter set for him. 'It's true. Siphiwe never broke tribal rulings.'

Ritter nodded, sliding into his own chair. Hosea leaned towards him. 'It was not our people who killed them. It was not Khumalo people, either. It was people from another place who want to make trouble between us, to start the fighting again. It is whites who do these killings, and put the blame on us.'

'Baba,' Ritter used the term of respect, Father: 'It's my job to find who killed your brother and your grandson. I'll do my job, but I need your help. Did Thulani say anything to you yesterday about going out with his uncle?'

'He told me he was going to my brother's house to help him write the labels for the medicine to be sent to the shop.'

'But they didn't write labels. They left the house.'

'I don't know why they did that. I heard nothing about it.'

'They were collecting medicines.' Ritter indicated two raw-hide pouches lying on his desk. 'Can you identify these, Mr Msomi? Don't touch them, please, just look at them.'

Hosea said without hesitation, 'The large bag belongs to my brother. The small one is Thulani's.'

'You're sure?'

'Yes, I know the markings.'

Ritter laid a plastic bag in front of Hosea. 'This was taken from your brother's bag. The stuff is still fresh, you see? The leaves are green and soft. They were collected last night.'

Hosea's face tightened. 'Perhaps Siphiwe went out for mutis, but he would not have set foot on Khumalo land. He must have been killed in another place.'

Ritter grunted. It was impossible to tell if he agreed, or not. Todd stretched out a hand.

'May I see, please?'

Ritter passed him the plastic bag. 'Don't open it, Mr Scallan. It's evidence.'

Todd studied the contents carefully, then said, 'They must have covered a lot of ground.' He indicated a root. 'This comes from marshy ground. The rest is from dry upland regions. Have you examined the gully where the bodies were found?'

Ritter nodded, frowning.

'Did the killings take place there?'

'I can't discuss that, Mr Scallan. We're working on it; we'll consider all the possibilities.'

'Was anything found in Thulani's pouch?'

Ritter hesitated a moment, then reached into a drawer and produced a second plastic bag, which he handed to Todd. It held a few large seeds, orange-red and shiny-fleshed. Traces of mast clung to them. Todd gave a soft exclamation and Ritter said sharply, 'Well, sir?'

'These are cycad seeds. Cycads grow everywhere on these hills, except on the east side of the village. The ground there's been stripped bare by crop-planters, or by overgrazing.

'So?'

Meeting the policeman's impassive stare, Todd felt a surge of anger. Thulani was dead, all that bright promise obliterated, but to Ritter it was probably run-of-the-mill stuff; just another case of tribal fighting.

'So Thulani didn't pick up the seeds on Khumalo land,' he said. 'I agree with Mr Msomi. The killing must have been done elsewhere, and the bodies dumped in the gully later last night.'

When Hosea and Ellen left the police station, Todd followed them out to the street. He felt the need to talk about Thulani, to express his own grief and try to share theirs; but Hosea made it plain they didn't want company.

'We have a lot to do, Mr Scallan,' he said. 'I must telephone my daughter, and talk with Siphiwe's family. The sergeant said . . . he told me we can't make arrangements for the funeral until the court gives us permission. We can't even bury our dead.'

'There has to be an inquest, you see? It's the law.'

'Yes. Well. Once that's done, we will ask Father Shabalala to make the service for us. In the mean time, people will be coming to our home. We must buy food and prepare it. We will be very busy.'

Seeing the hurt in Todd's face, Ellen said gently, 'You have been a kind friend to us, Mr Todd. Later perhaps you will come and see us, and we'll talk?'

There were black people gathered at the foot of the steps. Hosea and Ellen joined them and the group moved slowly away towards the post office.

Hearing a footfall behind him, Todd turned to see Ritter standing in the doorway. The policeman shook his head.

'A bad business. It could start the faction fights again.'

'You still think this was a tribal murder?'

'Well, sir, it's been going on a long time.'

'And what will you do about it, Sergeant? Will you give the Msomis the same attention as you are giving to the Smedleys?'

Ritter's expression became wooden. 'I'm not in charge of the Smedley case, meneer.'

'Who is?'

'Lieutenant Wolmarans, from Pomona.'

'I see. A lieutenant for the whites and a sergeant for the blacks.'

'No, meneer.' Ritter's voice was harsh. 'You must understand, I do my job the best I can. I think the Msomis were killed because they trespassed on Khumalo land. I don't know why the Smedleys were killed. I'd like to find out. I'd like to find out who killed all these people. Whether they black or white doesn't make any difference to me.'

He brushed past Todd, flung himself into the police car at the foot of the steps and drove off at speed.

Left alone, Todd made his way back to the hotel.

19

Max Duma arrived in Neil's office in a hurry.

'I'm due at a Peace Accord meeting in Thembisa,' he said. He produced the photograph Neil had given him. 'The man in the car is Frans Oswald. He was with the Bureau of State Security. Worked in the backroom evaluating information, keeping files on dangerous criminals like you and me. He specialized in illegal arms deals. Knew a lot we didn't want him to know.'

'Who told you this?'

'I have friends with long memories . . . and the scars to prove it.'

'You think Oswald was implicated in David's death?'

'There's nothing to show he was, but that pic was taken on the day of the inquest. Oswald was apparently interested in Silbermann. He must have known Wivern. He was a spider who knew other spiders.' Duma's teeth glinted. 'If he didn't kill anyone, he knew people who did.'

'Where is he now?'

Duma shrugged his heavy shoulders. 'I drew a blank there. He was retired three years ago. He was sick. Sick in the chest and sick in the head. That's the word.'

'Can you find out where he lives?'

'Maybe. There was an old woman worked for him, did the cooking and cleaning. She used to tip us off sometimes, when a raid was coming up.'

'Hedging her bets?'

'We all try to survive, man. Anyway, the old bag may still be working for him and sending money home to her family in Tzaneen. Perhaps we can find out something.'

He hurried off. Neil sat down at his desk and studied the photograph. The man at the car window stared toad-like at the retreating Silbermann.

Why had a backroom boy wanted to attend a public inquest? The men of BOSS had clung to anonymity. They had worked in deep shadow, sure they were right, convinced the end justified the means. No doubt Frans Oswald had seen David's death as necessary.

But if that were so – if David's death was what he sought, then he'd achieved his aim. There was no need for him to revive the case. There was no reason to think Oswald was in league with Wivern; no reason to connect him with the note.

Neil leaned his head in his hands. Nothing made sense. As he traversed the long corridors of memory, doors slammed shut before he could reach them. The truth constantly receded.

Outside in the street a heavy thumping began. He went to the window. Demolishers were attacking the building next door with a massive ball and chain. The noise was ear-splitting. Impossible to think with that row going on.

He picked up the phone and dialled Anna's number, not expecting her to answer, but she was home.

'That old photo album of David's,' Neil said. 'I don't suppose you still have it?'

She hesitated a moment, then said, 'Yes, I think I do.'

'May I come over and look at it?'

'Of course.'

When he reached her apartment, the album was on the

coffee table in the living room. Neil picked it up eagerly, carried it to the sofa, and began to turn the pages.

Anna said, 'What exactly are you looking for?' The careful neutrality of her tone made Neil look up.

'I thought I might get a lead on who took the photo of Oswald.'

She came to sit facing him. 'Who's Oswald?'

'I'm sorry, I'm not being coherent. The man in the photo you gave me, the man in the car – he's been identified as Frans Oswald, ex-BOSS agent.' He showed her the print. 'You see it's square, black and white, with deckle edges. That's very unusual these days. If we can find others like it in Dave's album, it would suggest . . .'

'. . . that one of Dave's friends took it?'

'Yes.' Neil turned more pages and came to a double spread of the same type of snapshot. Most showed groups of students, chatting outside the law faculty, building a Rag float, crowded together at a party. The same faces recurred. Their names, scrawled under the pictures, were familiar. David's boon companions.

'These last three at the party,' Anna said. 'Someone could have changed places with the photographer? If we can find one face that's different . . .'

They found a face that occurred in only one of the prints. The name beneath was Jiminy B.

'Jiminy B?' Anna said, 'James B? Can you recall a James B?'

'No, I can't place him.' Neil stared at the smiling features. He said slowly. 'Dave didn't bring all his friends home. Some of the most . . . significant . . . stayed away.'

'Significant in what sense?'

'Dave broke the law, getting people out of the country. Perhaps this Jiminy B worked with him. Perhaps they decided not to be seen together too often.'

'Wait.' Anna went away and returned with a cardboard folder. It contained a formal photograph of students in gowns, ranged in two tiers on a flight of steps. Anna's finger ran along the rows, stopped. She checked the name printed on the mount.

'Jerome Briony,' she said in triumph. 'Now do you remember?'

'I remember Ed Briony!' Neil stood up, excited. 'Briony, Kemp and Theunissen were attorneys. Damn it, they operated from the building next to the courthouse where the inquest was held! It's demolished now, but there's no question; the angle of the photograph is right. It must have been taken from Briony's office window.'

The firm of Briony, Kemp and Theunissen had literally gone up in the world. Their new premises were in a tower block close to the Central Court. A security guard directed Neil to the fourteenth floor and a receptionist ushered him with flattering speed into the office of Jerome Briony.

He came forward to shake Neil's hand; mid-thirties, plump, with crinkly black hair brushed back from a round pale face. Jiminy B, without a doubt.

'Mr Scallan, I'm honoured, sir. How may I help you?'

Neil settled himself in a swivel chair.

'I think you knew my son at university, Mr Briony. David Scallan.'

Briony's eyes became opaque. 'Yes. Yes, I did.'

'There's a matter I'd like to discuss with you. Perhaps you'll lunch with me?'

Briony made a show of studying his watch. 'I'm afraid I have a lunch appointment. If you'd care to tell me, here . . .'

Neil produced the photograph. 'I believe you took this.'

Briony glanced at it, leaned back in his chair, crossed his

legs nonchalantly; but the foot in its Gucci shoe twitched back and forth like an injured snake.

'Yes, I took it. An act of youthful bravado.'

'Why do you say that?'

'Well, in those days it was illegal to photograph any police business. As a pal of David's, I felt incensed. I took the picture.'

'Umh.' Neil turned the photo over. 'Is that your writing on the back?'

'Yes, it is.'

'Can you identify the man on the steps?'

'I'm afraid not. As I said, I acted on impulse.'

'His name is Colonel Hermanus Silbermann. He was in charge of the investigation into David's death. I visited him yesterday. He's in a retirement home. Quite senile, unable to communicate.' Neil saw this was not news to his host, who shrugged.

'One can hardly feel sorry for such a man.'

Neil put the photograph on Briony's desk. 'You were one of Dave's group, weren't you?'

'We were in the same year, both doing law. One horsed around with the gang, you know.'

The glibness angered Neil, and he said, 'Mr Briony, last week I received a phone call from a man calling himself George Wivern. Later that same day I received a note suggesting that Silbermann, Lemuel Moroka and Mark Alberton were paid large sums of money . . . presumably as payment for their work as a hit squad.'

Briony blew out his lips in a small puffing sound. 'Really, Mr Scallan, I don't see how any of this concerns me. It seems a matter for the police.'

'The police have been informed.'

'One hopes they'll be able to help you.' Briony seemed to realize he sounded callous, and he made a deprecatory grimace. 'I'm sorry. I know this must have been distressing for you.'

Neil brushed sympathy aside. 'Do you know why my son was murdered, Mr Briony?'

'He was an activist. He opposed apartheid.'

'Hardly a capital crime! I think you know David was one of a group helping political refugees across the border. I think you were part of the same group.'

As Briony opened his mouth to protest, Neil laid a deckle-edged black and white print beside the other photograph. 'This was taken at a party shortly before David died. The man sitting between you and David is "Molo" Thebahali, one of the people Dave helped to escape. Molo is back in this country now. He's prepared to testify to your membership of Dave's group.'

Briony was sweating. He wiped his fingers across his fore-head. 'He's lying. I don't have to listen to this crap.'

Neil watched him dispassionately. 'The amnesty covers you, you know. You needn't fear prosecution . . . particularly as you never seem to have come under suspicion. Mark Alberton lodged no information against you, you weren't questioned by the police.' He saw Briony's head jerk. 'Or am I wrong about that? Were you questioned?'

Briony burst out, 'What right have you to come here and accuse me? I don't owe you anything. You never concerned yourself with the rest of us. All you thought about was David.'

'You're right. I should have remembered his friends but, to be honest, I didn't know you figured among them. He never spoke of you. Did the police question you, Mr Briony?'

'Question?!' Briony was shaking violently. 'It went on for weeks, months. He never left me alone, he kept coming back . . .'

'Who did? Silbermann?'

'Wivern. He was like a cat, smiling, watching me, clawing me back every time I tried to get away.'

'He wanted the names of the others in the group?'

'Yes.' Briony lifted his chin, finding some shreds of dignity. 'I didn't tell anything. I said I didn't know about the group, I never helped them. Then one day . . . it stopped. Friends told me he'd been sent to Angola. I heard later he was killed in action. I thanked God. I went on my knees and thanked God.'

'It looks as if your gratitude was premature.'

'No! Wivern's dead.'

The words were spoken with utter conviction, and Neil changed tack.

'Do you remember Mark Alberton?'

'Of course I do. The little prick sold us out. Then he ran off to Australia.'

'According to the note, he went with seventeen thousand rand in his pocket. Do you think he made that much as an informer?'

'How should I know?'

'You'll be glad to hear he didn't live to enjoy it. He was killed a year later, in a car crash.' Neil watched Briony. 'So, of the investigatory team, three are dead . . . Wivern, Alberton, Moroka . . . and one is dying.'

'Good riddance. They fucked up all our lives.'

'They must have acted on orders.'

'Of course. The system . . .'

'I mean, a specific person in the security forces.'

Briony looked vague, almost listless. 'I don't know who controlled them,' he said. 'I don't want to know. It's over and done with. I just want to get on with my life.'

Neil picked up the photographs and rose to his feet. 'I understand, Mr Briony. Thank you for seeing me.'

He started for the door, but before he reached it, Briony

called after him. 'Wait. Please.' He came round his desk to where Neil stood.

'Mr Scallan, I'd help you if I could, but I don't know anything new.'

'I understand. I won't trouble you further.'

'I suppose you think I'm gutless.' There were tears in Briony's eyes. 'I was never like David. He had such a strong belief, it seemed to carry him through everything. I was afraid all the time. I never actually took anyone across the border. He made so many trips. He must have known that one day he'd be caught. I told him not to go to Gaborone, that last time, but he went. That was how he found out Alberton was working for the police.'

'What? How did he find out?'

'He saw Alberton in Gaborone, talking to a fair-haired man in a café. The refugee Dave was with said the man was a cop in the Special Branch in Jo'burg. When Dave came back from the trip he told me he thought Alberton was a paid informer. He said he was going to expose him; but the next day, Dave was killed. We – some of his pals – went to the university authorities and told them about Alberton. He was kicked out. You know about all that, though.'

'Did you ever identify the cop Alberton was with?'

'No, but I'm sure it was Wivern. I think Wivern knew he'd been seen, and came back here and set up the . . . the bomb.'

'To stop David blowing the whistle on Alberton?'

'Yes.'

'Do you think Alberton was important enough to merit murder?'

'He must have been. He was useful to them. They wanted to stop Dave's work. They thought they could do anything they liked. They thought they were God.' Briony's gaze lifted to Neil's face. 'I'm sorry I didn't tell you before; but after I

131

saw what they did to Dave, I was too scared. I was just . . . too scared.'

'You were brave enough to join the group,' Neil said. 'You didn't rat on your mates. You did more than most.'

Briony swallowed. 'I want to say . . . I want to help you. If you need me to testify, I will.' Uncertainly, he held out his hand.

Neil shook it. 'Thank you. As soon as I have things clear in my mind, I'll get in touch.'

Leaving Briony's office, Neil walked through the lunch-hour crowds to Joubert Park, found a bench and sat thinking about what he'd learned.

It shed light on David's last days.

Alberton must have been watching him for some time, must have tipped off the police about that final trip to Gaborone, must have travelled with a security policeman – Wivern, almost certainly – to the border town to complete their case against David. Having done so, they condemned him to death without benefit of trial.

Alberton had been far more than a small-time campus spy. He had been actively involved in covert action; an accomplice to David's murder.

Talking to Jerome Briony had brought Dave unbearably close. Almost he might emerge from the crowd at the gate; hurrying, scattering the pecking pigeons, vibrantly alive.

Neil groaned and pressed a hand to his eyes.

He remembered old Mr Alberton, his querulous voice saying, 'They paid his fees at varsity for a year, that was all.' Not 17,000 rand, then. What was lies, what truth?

He made his way back to his chambers, sat at his desk and wrote on a pad:

'Please extend your inquiries to the death of Mark

Alberton in a car accident in the Brisbane area of Australia, probably in late 1984 or early 1985.'

He signed his name to the message, went to the fax machine and dialled Louis Baird's number in Rome. The signal to proceed came up, and he fed the paper into the machine.

20

In his room at the hotel, Todd picked up the rucksack Thulani had always used, and stowed it in a cupboard. His throat felt tight with anger and grief.

His window overlooked the garden at the back of the building. Beyond that was a lane, a stream, then the steep side of the hill. A line of people was moving slowly up the path towards the Msomi kraal. When they reached Siphiwe's home, Hosea's home, they would enter the living room and sit awhile with the mourning family. They would hand over small gifts of money to help pay the expenses of the coming funeral. They would offer their speech or their silence. They would make themselves part of a death that concerned their whole community.

Any man's death diminished them, because they were involved in mankind; as I am not, thought Todd. He remembered his father's angry words, 'You always try to avoid unpleasantness.' It was true. Like his mother, he had turned away from the horror of Dave's death, from his father's efforts to track the killers, from that latest note from Wivern.

It didn't matter that in all likelihood the note was a fraud. It was part of the evil that possessed society; part of the fragmentation, the loss of conviction, the fear that led to a fatal inertia of the conscience.

It was easier to ignore the truth than to face it, and so the decay spread. Dave's death was only one in a chain of many thousands. Thulani's death was the most recent link in that chain, but it wouldn't be the last. The killing would go on until the thousands of living moved to stop it. One thing was necessary, the power of the individual conscience.

He stretched out a hand to the telephone, then drew back. He couldn't speak to Neil on this line. The girl on the switchboard was the village gossip.

Most of the village people used party lines. Lift one receiver, and you heard the clicks as the eavesdroppers got set to listen.

He would have to drive over to Pomona.

As he passed the reception desk, he saw a man signing the register. He was fair-skinned and fair-haired. He gave Todd an indifferent glance. His eyes were set wide apart, turquoise blue. Leaning against the desk was a rifle fitted with a telescopic sight.

Todd left the hotel by the back door, climbed into his station-wagon and headed for du Toit's store.

'The boss is out the back.'

The Indian at the counter looked harassed, dealing with a group of labourers. Todd walked round to the yard behind the store and found the owner tinkering with the engine of his delivery van. He straightened up as Todd approached.

'Morning, meneer. What can I do for you?'

'I want to buy a gun.'

'Ja.' Du Toit wiped big hands on a piece of cotton waste and tossed it aside. 'I heard your boy was killed last night. You wise to buy a gun, but I can't sell you one. Don't carry that kind of stock.'

'Is there a gunsmith in Pomona?'

'Ja, right opposite the hotel.' Du Toit screwed up his eyes, studying Todd. 'You had that boy long?'

'He worked for me on holidays, for three years.'

'Umh. I'm sorry. That's a good family, the Msomis.'

'Sergeant Ritter thinks they were killed because they tres-passed on Khumalo ground.'

'It's happened before. There's been fighting in those hills, long as I can remember.'

'Hosea Msomi says there was a truce. He said "Siphiwe would never have moved off our land."'

Du Toit pulled a stubby pipe from a pocket and began to fill it. 'So what do you think happened?'

'I think they were killed elsewhere and dumped in the gully by the killers. Maybe it happened to the north, on the Smedleys' farm.' Todd was thinking of the forest-clad stretch along the farm boundary; but du Toit shook his head.

'The cops are watching that side of the valley.'

'Four murders in a week? There must be a connection.' Du Toit shook his head again, but Todd persisted. 'You were friendly with Smedley. Did you see him the night he was murdered?'

Piet frowned, sucking at his pipe. He came to a decision. 'Ja, I did. I saw him about seven o'clock. I'd been to Pomona to fetch supplies. About five kays from here, near where Smedley's farmtrack joins the hard road, I saw a closed van pulled into the verge. It was about a hundred yards ahead of me. Smedley's car was stopped opposite it. I couldn't get past, so I waited.

'Smedley was talking to a black man who was changing a rear tyre. He got out and shone his torch on the man. Then he went to the front of the van and shone the torch into the cab. He stood there for a minute or so talking. Then he climbed back in his car and came towards me. He stopped when he drew level, asked where I'd been. He was a nosy old bastard, you know? He told me he asked the people in the van if they needed help, and they said no. I got the feeling they told the old man to bug off, because he said – kind of kidding – "That van must be full of dagga, the way they acted."

'I asked which farm they were looking for, and Smedley didn't answer. He was staring back at the van, as if he was trying to remember something, know what I mean? Then he drove off up the track. That was the last time I saw him alive.'

'Do you think he recognized someone in the van?'

'That I can't say, meneer.'

'Did you see them yourself?'

'I took a look as I passed. The black was round the far side of the van, and the ou in the cab was lying back with his hat over his face.'

'Did you tell Ritter about this?'

'Ja, I did. He asked me, what was the van like? I told him, plain green with dirty number plates.'

'What's your opinion of Ritter?'

Du Toit gave a faint shrug, puffing at his pipe. 'Man, I hardly know 'im. He's only been here six months.'

'Do you trust him?'

'Trust?' Du Toit gave a sudden snort of laughter. 'I don't trust anyone. I'm a shopkeeper.' His face became more serious. 'I trust my boerbull and my gun. Get yourself a dog and a gun, Mr Scallan, that's my advice to you.'

Pomona lay on the dry side of the barrier of hills, thirty streets long and twenty wide, three each way given over to shops, service stations, three churches and a town hall.

The Pomona Hotel lay dead centre, a two-storey Victorian building with brookie-lace ironwork along its verandas, and its name in white on a red tin roof.

In the foyer Todd obtained coins from a desk clerk and was directed to the callbox at the end of a passage.

He dialled Neil's office number and was told Mr Scallan was out, no one knew where. He tried his own home and got Aline.

'Todd?' She sounded confused. 'Darling, what is it, is something wrong? Your voice sounds funny.'

'Yes.' He sought for words, and at last blurted out, 'Thulani and his uncle Siphiwe were murdered last night.'

'Murdered? Did you say . . . ? Todd?'

'They were hacked to death.'

'But . . . why? Was there fighting?'

'The police say there was. I don't agree. Ally? Don't tell Joe for the moment.'

'Oh, no. Oh, poor Thulani.'

'Do you know where Dad is? He's not at his office. Did he say where he was going?'

'No.'

He made a decision. 'I'm coming home. I have to talk to him.'

'When will you come?'

'I'll leave here early tomorrow. Home by nine. Tell Dad he must stay home, it's important.'

'All right. Todd, please be careful.'

'I will. Love to everyone. See you tomorrow.'

The gunsmith's shop was, as Piet du Toit had said, right opposite the hotel, wedged between the supermarket and a hardware store. Heavy iron grilles coated with dust covered its twin display windows. The door was locked and Todd had to wait for a wizened old proprietor to open it.

It was clear that the old man dealt more in maintenance than in sales; but after some delay he unlocked his gun safe and produced an army rifle of recent make.

'You know how to handle this?' he demanded.

'Yes.' Todd took it, examined it, laid it on the counter. 'It'll do. The price is OK. I'll need ammo.'

'Just like that?' The old man was scandalized. 'You should test it, first!'

'Why? Is it defective?'

'No, course it isn't. You think I'm a crook?'

'No, you have an honest face.'

'Huh!' The old man edged round to the far side of the counter. 'Why d'you want a gun? You a farmer?'

'No, a botanist, I work in the field.' Todd hesitated. 'My assistant was killed last night.'

'Ah, yeah, then you'll be glad of this.'

Todd handed over his credit card. The old man closed the deal and produced a battered leather gun case.

'Free with every purchase,' he said. 'You'll need a permit, you know.'

'Yes. Where's the police station?'

'Five along from here.' The old man followed Todd to the door and watched him walk along the street.

'Just like that! Treat guns like liquorice sticks, these days. Bloody gangsters, the lot of 'em.'

Todd arranged for his permit and went back to his car. A four-wheel drive was parked next to it. Its driver raised a hand in salute. Blue cat-eyes glanced at the case Todd carried, then shifted to his face.

'Shopping?' The voice was flat and faintly mocking.

'That's right.' Todd unlocked the car door, laid his purchase on the back seat.

'Dangerous,' the stranger observed. 'Anyone can hop in, blow your head off with your own gun.'

'You seem experienced.'

'I was in the army for twenty years. People kept trying to shoot me.'

'You're staying at the Muller's Drift Hotel?'

'Not for long, I hope.'

'See you, then.'

The fair man nodded and flicked the key of the four-wheel drive. A bright cloud of dust marked his progress along the street.

The cloud remained an even distance ahead of Todd all the way back to Muller's Drift. It gave him the odd sensation of being followed. He lifted the rifle from behind him and placed it on the passenger seat.

As soon as he could, Todd found a chance to examine the hotel register. The last name entered was Leonard Brown. The address alongside it was in Braamfontein. Todd doubted it was genuine. He felt instinctively that the man was a liar, a predator who never gave up on his prey.

At lunch, Brown chose a table near the door. Once, Todd looked up to see the light eyes watching him. Their expression was bored, almost blank. Deliberately, Todd stared back and the fair man turned his head away.

He had a very distinctive face, Todd thought. If he could get a photograph of it, he might be able to discover Brown's real identity.

Finishing his meal, he went to his room and collected his video camera. Returning, he saw Brown sitting in one of the easy chairs in the lobby. Todd walked past him without a glance, turned out of the back door and hurried along the path to the corner of the building.

A hedge of plumbago separated the path from the flower garden, and beyond the hedge grew an massive old thunder tree. Todd set the camera going, levelled it at the tree, and waited.

He heard footsteps hurrying from the back door. Brown appeared at the corner and stopped as Todd swung the camera down, holding it steady for a moment before lowering it.

'You bloody fool,' he shouted. 'I had an oriole in frame, up there!'

'I'm sorry.' The fair man gazed up at the tree. 'Was it a black-head or a golden?'

The answer was so unexpected that Todd hesitated. 'You know birds?' he said.

'Most. I've spent a lot of time in the veld. Watching birds is something to pass the time.'

Brown took a half-step back; placatory, retreating from an angry rival's territory. 'I heard you lost your porter last night,' he said.

'Yes.'

'Are you looking for a substitute? If so, I'd like to apply for the job.'

He seemed serious, his voice was almost humble. Todd shook his head.

'It can't be called a job. I gave Thulani a holiday task, paid him by buying his school books.'

'All I want is enough to cover my board and lodging. I know the hill country. I speak good Zulu. I can use a gun.'

Brown was smiling, as if they enjoyed a shared joke. It came to Todd that what he was being offered was protection. He opened his mouth to refuse the offer, then changed his mind.

He could not hope to discover who killed Thulani by hovering on the outskirts of things. He had to reach the centre. Brown might be a way in.

'Come inside,' he said, 'and we'll work something out.'

21

Colonel Marx lit one of the Dutch cheroots he favoured and fixed Joubert with a cold stare.

'So why am I here?'

'Scallan was in this afternoon. He asked a lot of questions.'

'Such as?'

'Why was Wivern sent to Angola; was he pleased with the transfer; was he sent to an army unit and, if so, why; how come he was standing near a petrol dump when it blew; who witnessed the incident; who testified to his death?'

'And you said?'

'I read him the official report.' Joubert touched a file beside him. 'The dump was hit by enemy aircraft. Poom! Georgie Wivern, dressed in sashes, fell in the fire and was burned to ashes. The ashes came home in a jar. The court said the man was dead. Scallan said, perhaps it was another man who died, perhaps Wivern escaped, went abroad, started a new life. I said, not possible. Lieutenant L.R. Bredenkamp of the tank corps saw Wivern burn.'

'What did Scallan say to that?'

'Nothing. He changed the subject. He said he'd been reliably informed that his son David saw George Wivern and Mark Alberton in Gaborone, the day before his death.'

Marx removed the cheroot from his mouth and laid it on an ashtray.

'That's news to me. It's not on the record.'

'News to everyone, sir. Silbermann's report specifically states that Alberton was at the university, and Wivern was in Jo'burg, for the whole of the week before David Scallan's murder. He gave the same testimony at the inquest.'

'Who was Scallan's informant?'

'He wouldn't say. We can pressure him.'

'No. I want to keep him friendly.' Marx spoke abstractedly, searching some recess of his mind.

'He asked for photos,' Joubert said. 'Silbermann, Wivern, Moroka and Alberton. I gave them to him. He could have applied elsewhere, he's a determined bugger. He claims he's identified the man in the snapshot. Says it's a security cop by the name of Frans Oswald.'

Marx roused from his reverie. 'Ex-security cop,' he said, and caught Joubert's look of surprise. 'I did some checking. Oswald was with BOSS for years. When that folded he moved to MI. His expertise was the illegal supply and distribution of weapons. He knew who sold them, how they came into the country, who bought them, distributed them, used them. Like all experts, he couldn't see beyond his own field. He didn't like the military. Especially, he didn't like the dirty tricks brigade. Early in the eighties, he started sending in reports about "unauthorized activities". He stamped on some important toes. He was reprimanded, then side-lined. Three years ago he was retired on psychiatric grounds. "Unbalanced," the report said, "an obsessive neurotic." One of his obsessions was that George Wivern faked his death.'

Startled, Joubert said, 'Is that on file?'

'No.' Marx smiled bleakly. 'Oswald's file is a wee bit thin. When I asked about him, I was told, he's a nut case, he's history, don't waste your time. Last night I had a drink with an old friend. He said, these days the Service leaks like a sieve. Oswald could still be getting information. He could still be active, on a limited scale.'

'A crazy who thinks Wivern's alive? He could have pulled the trick on Scallan.'

'True, but that's not our main problem right now.' Marx waved aside a feather of smoke. 'Our problem is, if Silbermann lied in his report, and at the inquest – if Wivern and Alberton were in Gaborone the day before young Scallan died – then it supports the theory that they were implicated in his death. The case must be re-examined.'

'Who was Oswald's chief?'

'At the time he was fired, Fernhutter.'

'He's retired. We'll have to approach someone else in the department.'

'Slowly, slowly now. We have to be careful, Tiaan. The person or persons who blocked Oswald's reports – who got him discredited and kicked out – may still be around.'

'What about Oswald himself? Where's he?'

'Gone to ground.' Marx stubbed out the butt of his cheroot. 'I'm looking for him, I'll find him, but until I do, I want you to keep your mouth shut about this. Understand?'

Neil arrived home late for dinner. Meg came to the door of the dining room, table napkin in hand.

'Where the hell have you been? You left the office without telling your secretary. Todd's been trying to get hold of you. He's worried about something, he's coming home.'

'Good, when?'

She came towards him, barring his way. 'It's not good. I demand to know what's going on. What are you up to?'

'Nothing.' He saw that she was on the verge of one of her hysterical attacks. 'Nothing to worry about, really.'

'Don't lie to me! You never tell me the truth. You only tell that whore of yours.'

'Meg, please, don't distress yourself.'

'Distress myself? What do you expect? I know you confide

in her, all the time. You were with her on Sunday at the exhibition; everyone saw you. Of course, Aline's on your side, she tried to make out it was just chance, but I know better; I saw through that little tale.' She snatched at his arm. 'What's happening? I have a right to know.'

'I assure you it has nothing to do with Anna.'

'Then what? What? Do you think I'm a child? Don't treat me like a child, Neil. Tell me the truth. You tell me. Now!'

Her grip on his arm was vicious; she twisted his flesh back and forth, shouting at him, 'You tell me, you tell me now!' He thrust her away, filled with a sick hate of her.

'Very well, I'll tell you. On Friday, a man phoned me claiming to be George Wivern. He's obviously a faker, but I have to sort it out. I spoke to the police . . .'

'And to her, of course! You spoke to her, not to me!'

'I asked Anna to check something in the university files. It was nothing personal . . .'

'Nothing personal? David was my son. You went to her about David and you never told me!'

'Meg, I wanted to save you pain. The call was probably a hoax. We'll find the hoaxer and put an end to it.'

'We? Who's "we"? You and your tart?' She pressed her hands to her head, glaring at him. 'That's all you ever cared about, fucking your whore. If we'd had a proper home, he'd never have mixed with that black trash, he'd be alive. He'd be alive.'

'Stop that!' Neil caught hold of her shoulders. 'Meg, stop that at once, do you hear?'

She began to fight him, clawing at his face, trying to sink her teeth into his hand. Her screams brought Aline and Philomela running. They pulled her away, took her shrieking and struggling up the stairs.

Neil went to the study and called her doctor. He came within ten minutes, hurried to her room and gave her an

145

injection. Neil remained downstairs. When the doctor had left, Aline came to stand beside his chair, bent to kiss his forehead.

'It's OK, she's asleep.'

'Thanks, Ally.'

'Dr Knowles is sending in a night nurse.'

'Good.'

'She found out Anna was at my exhibition. Some fool woman told her. I told her it was nothing.'

'She wanted to know the truth,' Neil said. 'I told her about the call from . . . Wivern, whoever it was. It set her off.'

'She was calling for David.'

Neil looked up wearily. 'It's not David she's concerned about, it's herself. Nothing must threaten her little life.' He knew that to be true, at last. Meg had never shared his pain about David's death. She had put him out of her mind, systematically filled his space with trivial things. The years spent trying to shield her, to comfort her, had been wasted. Facing that fact, he felt a sense of release.

'You'd better go, my dear. Joe will be wondering where you are.'

'I'll come back later to see how things are.'

'They'll be all right.'

When she'd gone back to her side of the house, he stayed in his chair. He heard the night nurse arrive. Philomela took her upstairs, and then brought him a tray of food.

He made no attempt to eat. Meg's tirade had momentarily drained him of energy; but as he sat quietly, the confusion that had clouded his mind lifted a little.

Dave had worked for nearly two years helping political refugees to escape. Mark Alberton had spied on him, and must have accumulated enough evidence to justify an arrest. Yet Dave had not been arrested.

The security police of that time would have relished a

chance to charge and try the son of Advocate Neil Scallan. They had been cheated of that chance.

David had been killed to ensure his silence. He had been killed, not because Wivern and Alberton saw him in Gaborone, but because he saw them.

Neil came out of the house to meet Todd.

'We'll talk in the garden, if you don't mind. Meg had one of her attacks last night. The doctor gave her an injection, she's still asleep.'

They crossed the lawn to the tennis court and sat on a bench in the shade. Neil studied Todd's face.

'I'm sorry about Thulani. He was a fine boy.'

'I owe you an apology,' Todd said. 'I should have backed you from the start.'

'Tell me about the Msomis.'

Todd described the interview at the police station, Ritter's theory that it was a tribal killing committed on Khumalo land.

'I don't believe it,' he said. 'Hosea told Ritter categorically that Siphiwe would never have trespassed, that he would have collected his medicines on their own land.'

'Where is that?'

'Their kraal is north of the village.'

'If the Khumalos didn't kill them,' Neil said, 'who did?'

'The same people who killed Bert Smedley and his wife, last week.' As Neil looked up in surprise, Todd insisted. 'The Smedleys weren't murdered for gain. They were broke. They'd sold the farm and all their stuff, and were leaving to come up here, to their daughter. It wasn't a vengeance killing, either. They were well-liked by everyone, black and white. Good employers, good neighbours.'

'So?'

'Yesterday, I talked to Pieter du Toit. He keeps the local

store, knows all the village gossip. He told me that on the evening of Smedley's death, he saw the old boy nosing round a van stopped at the side of the road, a few kays from the village. The driver was changing a flat tyre. Smedley offered help, which the driver refused. Then Smedley tried to chat to the passenger in the cab. He was snubbed. He told du Toit that the people must be smuggling dagga, the way they saw him off. Maybe he cracked that same joke to them, maybe they took him seriously.'

'You think someone's running pot out of Muller's Drift?'

'No. The area doesn't provide enough cover for dagga fields. The white farmers use every inch of their arable land. The black reserves are over-grazed and barren as hell, you could spot a dagga patch from way off. But something's going on that provides a motive for murder. Smedley recognized someone in that van . . . or someone recognized him. He stuck his nose in where it wasn't wanted and a few hours later he and his wife were dead. Thulani and Siphiwe went on a moonlight walk on the hills – the same side of the village as the Smedleys' farm – and they're dead.' Todd turned to face his father. 'I think the passenger in the van was George Wivern.'

'There's nothing to link Wivern with what happened in Muller's Drift. No common factor.'

'We're the common factor. The Scallans. Dave, you, me; we're what ties this all together. We're the target.'

Neil moved uneasily on the bench. 'Todd, listen. In this kind of situation, it's easy to become paranoid . . .'

'I am not paranoid. There's something you don't know. A man arrived at the hotel yesterday, followed me round, asked if he could take over Thulani's job. In fact, he was offering to be my watchdog.'

Neil stared, then said abruptly, 'Describe him.'

'Fortyish, medium height, fair hair worn long, fair skin

with a lot of tan, blue-green eyes set wide apart. In good physical condition. He said he'd been in the army for twenty years.'

'What name did he give?'

'Leonard Brown.'

Neil reached into his inner pocket and drew out the photos Joubert had given him. He passed one to Todd.

'Recognize him?'

'Of course. It's Wivern.'

'Not Leonard Brown?'

'No, nothing like, except for the light-coloured hair. I got him on video. You can show the cops, check him out.'

'The cops don't keep records of soldiers.'

'I think he may have lied about his army service. I think he's a liar. He never expected me to believe him. He almost made a joke of asking for the job.' Todd paused, then went on. 'I think he's dangerous. I'd like to know who he works for.'

'A man called Frans Oswald, I think.'

As Todd looked inquiring, Neil told him about Max Duma's identifying the man in the snapshot, and his own subsequent interview with Jerome Briony.

'It suggests a new motive for David's murder,' he said. 'Dave saw Wivern and Alberton in Gaborone. He was going to report it. He was killed to prevent him from doing so.'

'Briony never spoke?'

'He was afraid to. They got Dave, he could have been next. He'll testify if we need him. Meanwhile we have a lot to do. Make a copy of your video and I'll give it to Joubert. He may be able to establish Brown's real name. Also, we must try to find out if Smedley ever knew Wivern.' Neil stood up. 'I'll phone Joubert and ask if we can meet him. We'll draw up a list of points we want to raise with him.'

Todd was thinking of something else. 'Last time I was here,' he said, 'you asked me to find out where Ritter was stationed before he came to Muller's Drift. Why was that?'

'You remember the bus massacre in Vryheid, about three years ago?'

'Yes.'

'Ritter was involved in a shoot-out with the attackers. He was badly injured and landed in hospital. An unrest monitor interviewed him and heard him mention a Mac Eleven – a rapid-fire machine-pistol.'

'Guns,' Todd muttered. 'It always comes back to guns.' He looked at Neil. 'Ask Joubert if Ritter's sent in any reports lately about the situation in Muller's Drift.'

22

'My name is Marx, Mr Scallan. I'm in charge of this department. And this is . . . ?'

'My son Todd.' Neil shook hands with the small man. Jewish, he thought, clever, hard-headed. 'Thank you for giving us your time.'

'Major Joubert thought I should meet you.' Marx waved the visitors to chairs, chose one himself. He gave Neil a frosty smile. 'So what have you got for us today?'

'A possible scenario,' Neil said.

Marx spread his hands inviting speech, and Neil began: 'Last Saturday I lodged a complaint with Major Joubert. It concerned a note purporting to come from ex-police officer George Wivern. I hoped for a police investigation. That hasn't happened.'

'Certain inquiries have been made,' Marx said, 'by Major Joubert and by myself.'

'Without result. I feel my complaint has not been taken seriously.'

'You must remember that no serious crime has been committed.'

'I believe that the note is associated with major crimes, Colonel Marx.'

'Indeed? Do you have anything to support that theory?'

'I said I have a scenario to present. It might link a little undisputed fact to a great deal of supposition. I think it would pay you to hear me out.'

Marx shrugged and leaned back in his chair.

'On a subsequent visit,' Neil continued, 'I asked Major Joubert if the name MacAllivan meant anything to him. He checked his computer data and couldn't find any mention of that name but when I told him it might be connected with guns, he suggested it referred to a Mac Eleven rapid-fire machine-pistol.'

Marx nodded. 'I believe,' he said blandly, 'that you couldn't recall where you heard the name?'

'I read it in a report. The man who mentioned it was a Sergeant C.A. Ritter – the hero of an incident in Vryheid three years ago.'

As Marx looked blank, Joubert said quietly, 'White Torch gunmen shot up a bus rank, killed a lot of blacks. Ritter and his partner went after the killers. Ritter was badly wounded. Right, Mr Scallan?'

'Right. While he was in hospital, Ritter was interviewed by unrest monitors. He described one of the bus rank killers as "a crazy who collects guns." Then he muttered something which the monitor recorded as "MacAllivan or Machiavellian". One may infer he suspected the White Torch arsenal could include a Mac Eleven.'

'If the sergeant had any such suspicion, he would have put it in his official report,' said Marx flatly.

'Precisely. It's a point that can be checked, can't it? One would also like to know if the White Torch gang had a cache of arms and, if so, where they obtained them.'

Marx made a note on a pad. 'How does this fit into your scenario, Mr Scallan?'

'My theme is the illegal arms trade. We all know it exists. Arms are smuggled in and distributed to the best markets. One such market is in the midland and northern areas of Natal. Would you agree?'

'Yes.'

'Good. To return to Sergeant Ritter – my information is that he was commended by the Minister of Police for his bravery, but did not, as one would expect, receive promotion. Instead, he was transferred to a backwater in the Northern Cape. Six months ago he was moved again, this time to Muller's Drift in Natal – another backwater. It's almost as if someone wanted him out of the way.'

'Are you suggesting a cover-up, meneer?'

'That would be one explanation. Another would be that Ritter was set to watch the routes used by gun-runners. I'd like to know if he's lodged reports since he was shifted to Natal, if those reports concern arms, and if any action has been taken on them.'

'I have a question for you, Mr Scallan. Why are we discussing the arms trade? What has that to do with your original complaint to the major here?'

'I received a note,' Neil said, 'listing the names of three men who with Wivern investigated the murder of my son. The note suggested that each of these men received a substantial sum of money. My immediate reaction – biased, I admit – was that they were paid as members of a hit squad. I now believe they were common criminals, bringing illicit arms into the country and arranging for their sale to the highest bidder.'

'It's a serious matter, meneer, to make unfounded allegations of that nature.'

'I said this was a scenario, remember? I do have some support for my theory. The week he was killed, my son travelled to Gaborone with a political refugee. While there, he saw George Wivern and Mark Alberton talking together in a café. They appeared to be on familiar terms.'

'It's common cause that Alberton was an informer.'

'He could have done his informing in Jo'burg. Why go all the way to Gaborone unless he and Wivern had other business in hand? At the inquest, Silbermann stated on oath that both those men were in Johannesburg for the whole of the week before David died. Why would he lie, unless it was to conceal criminal activity? I believe David was killed because knowledge of what he saw would be dangerous to Wivern and his gang . . . and to people a great deal higher in rank than any of them. I would like to know who those people were.'

Marx's fingers rose and fell on the desk top. 'Mr Scallan, we have special squads whose sole duty is to keep a stringent watch on arms movements. We work round the clock to prevent illicit weapons coming into the country.'

'Yet they come in. I spoke to a friend, the other night – a very influential friend – who says that power groups in South Africa are buying new sophisticated weaponry, the sort of things that can be used in urban warfare.'

'What are you suggesting?'

'That Wivern's group was bringing in arms, facilitating their sale here. As policemen they could do things that are difficult for civilians. They would have advance warning of police or military action aimed at halting the trade. I'm also suggesting that some of the top operators in that trade are still in place.' Neil raised a clenched fist. 'Against that background, Colonel, you have to consider the idea that George Wivern is still alive, that he's controlling some branch of the operation, that he has perhaps been seen by someone strongly opposed to that operation, someone who hates Wivern and regards him as a traitor, but does not have the power to stop his activities. That person arranged for me to receive the phone call and the note.'

'Can you put a name to this mystery man?'

'Yes. Frans Oswald, the man in the photograph.'

Marx was silent for a moment, running a hand along his jaw. Then he said, 'If Oswald had information about Wivern, why didn't he come to the proper authorities? Why did he approach you?'

'Because he's down on your records as being mentally unsound. Because he knew his report would be dismissed as the ravings of a lunatic. Because he knew that I would act on it, take it to the Press or the police. Either way, some action would result.'

Marx sighed. 'Mr Scallan, Frans Oswald is mentally disturbed, I know that for a fact. What do you expect me to do?'

'Test my theory, accept the scenario that Wivern was involved in the gun trade, that he's still alive. Check your records, look for reports from Oswald, from Ritter.' As Marx seemed about to rise from his chair, Neil put a hand on his arm. 'Wait, Colonel, I'm not done yet. Guns come into this country by several routes: across the northern borders, Zimbabwe, Mozambique – through Zululand, Richards Bay. It's necessary to stash them somewhere. It's possible that Muller's Drift is one of the points where they're hidden. My son Todd has something to say about that.'

Both Marx and Joubert swivelled round to stare at Todd. He swallowed nervously, then said, 'At first, I thought the note was a fake. Now I don't. Too many people have been killed.'

Marx began an angry movement, checked it. 'What people do you mean, Mr Scallan?'

'My brother David,' Todd said. 'Alberton and Moroka, Mr and Mrs Smedley and Thulani and Siphiwe Msomi.'

'You think these deaths are connected?'

'Yes. To Wivern.'

'Why do you think that?'

'Everything fits.'

Marx made a sound half-angry, half-amused. 'Sir, I could invent a hundred scenarios that would "fit", as you call it.

That wouldn't make them true. You have no proof that Wivern's alive.'

'There could be a way to prove it.'

'How?'

'Mr Smedley – the farmer who was murdered in Muller's Drift – saw someone in an unmarked van, the evening of his death. Someone he recognized, perhaps.'

'And you're saying that "someone" was Wivern?'

Todd ignored Marx's derisive tone. 'If we could establish that Smedley knew Wivern? You could ask around, couldn't you? The police can ask questions. Find people.'

'Where do you suggest we start?'

'Smedley's daughter lives in Johannesburg. Smedley and his wife were going to live with her and her husband.'

'Do you know the lady's married name?'

'No . . . I'm afraid not.' Todd sounded humble, and Marx looked at him curiously.

Turning to Joubert, Marx said, 'Check up on it. Anything known about Smedley's past associates.' He turned back to Todd. 'Anything else?'

'Yes.' Todd laid a video cassette on the desk. 'A man signed in at my hotel in Muller's Drift. He followed me around, asked for a job. I don't trust him. I'd like him identified. He calls himself Leonard Brown, but I don't think that's his real name.'

Marx opened his mouth, shut it again, handed the cassette to Joubert. 'Run it, Tiaan.'

Joubert slotted the tape into the machine, started it running. A picture of a plumbago bush appeared on the screen. Then the camera veered wildly, coming to rest on the figure of a man.

'Stop there,' Todd ordered.

The figure was frozen in mid-stride, the face stared directly at them, wide-eyed and fierce. For a moment no one spoke, then Marx said, 'Did you give him the job, Mr Scallan?'

'Yes I did. Do you recognize him?'

Marx shook his head; denying knowledge, or refusing an answer? 'I'll have him checked,' he said. 'Can I keep the tape?'

'Sure.'

'Send it over to Records,' Marx told Joubert, 'and you'd better see if Ritter knows the man.'

'No,' said Todd sharply, 'don't say anything to Ritter. Not yet.'

Marx regarded Todd as a man regards a strange fish washed up on his beach. 'Why, meneer?'

'Ritter may be the salt of the earth, but he's no diplomat. If he blunders in and questions Brown, he could blow the whole thing. You have to find Oswald and question him; you have to deal with the questions my father's raised; you have to keep watch on Muller's Drift. You mustn't alert Wivern. Not till we're ready.'

'Ready for what?'

'Ready to trap him.'

Marx drew a long breath. 'Sir, you don't begin to understand how a police department works. We can't spend the public's money on . . . on . . .'

Neil intervened. 'We appreciate your position, Colonel. If you think we've been talking rubbish, you should forget the whole thing. The decision's yours. Come along, Todd.' He started to rise, but this time it was Marx who put out a restraining hand.

'Sit down, Mr Scallan. Please. Sit down.'

Neil sat. Marx steepled his fingers. 'Who told you about Frans Oswald?' he said.

'A friend who has good reason to remember him.'

'Did this friend tell you about Oswald's work?'

'He said he worked with BOSS, and later with Military

157

Intelligence, in the section dealing with illegal arms deals.'

'Correct. Specifically, his task was to collect and collate information about people engaged in smuggling arms into South Africa. He was a brilliant operator . . . a man with a mission that became an obsession. He believed in the total onslaught . . . a world full of enemies bent on overthrowing our government. He particularly hated gun-runners. It was a deeply personal hatred, you understand? He saw it as his God-given task to break the trade.

'He pursued that task with fanatical devotion. He was not liked by his colleagues. They treated him badly. He became first resentful, then irrational. He began to accuse people in high places of being party to illegal arms deals.'

'What high places?'

Marx gave a twisted smile. 'Armscor, the military, the police, you name it. Of course, such behaviour couldn't be tolerated.'

'He was fired?'

'Side-lined, rather. He was buried in a stuffy little office a long way from the centre of things. His job was to deal with minor infringements, overlarge expense claims, overspending on trips abroad, that kind of petty *verneukery.*'

'What about payments to hit squads?'

'No. That sort of information − if it existed − would be classified and quite beyond his reach.'

Neil remembered the photograph of Oswald; watching Silbermann with malign absorption.

'He wouldn't have accepted defeat,' he said. 'Given his convictions − his obsession, if you like − he would have continued to probe for evidence of gun-running. Perhaps he found proof of money deposited in a foreign bank? Something that led him to suspect Wivern and the rest?'

'It's a possible scenario.'

'Did Oswald ever make accusations against Wivern?'

'I'm told he did. He made accusations against so many people. In 1984, he took the view that Wivern had not been killed in action, but had faked his death, and was still alive.'

'Are these statements on record?'

'No.' Marx spoke with deliberation. 'There is nothing on record that relates to Oswald's statements, accusations or actions with regard to Wivern.'

'That seems extraordinary.'

Marx gave a faint shrug. 'He was regarded as a crackpot. When he cried wolf, no one listened.'

Neil met Marx's cold grey eyes. 'If no records exist, Colonel, how come you know what happened?'

'Let's just say I also have friends with good reason to remember,' Marx replied.

23

'That was a waste of time,' Todd said, as they left the building. 'Marx as good as told us we have no case, nothing to investigate, no evidence that any crime's been committed against us.'

'He listened. He's a clever man and a good policeman. He knows that in times like these the laws of routine procedure don't work. There's too much alienation within the old structures. When a regime crumbles, the only way to survive is to act on one's own initiative.'

'You think he'll help us?'

'I think he'll try. We'll have to wait and see.'

When Joe came home from school at lunchtime, Todd told him about Thulani's death. The boy clung to his father.

'I don't believe he's dead. I don't want him to be dead, he's my friend.'

'I know.' Todd picked the child up and carried him to a chair, sat there holding him. 'It's very hard to bear when our friends die.'

'I want them to catch the people who did it and put them in prison for ever.'

'We're trying to find them, Joe. Meanwhile we have to remember Thulani the way he was, a good mate.'

Joe rubbed a hand across his eyes. 'He knew a lot of things. About football. He knew the names of all the players.'

'He did, yes.'

'Last time he was here, he was telling me about grasses and stuff like that. He wanted to give you a present, because you helped him at school. He was looking at your books. He said he had an idea for a present.'

'It was a good idea. A man can't have too many books.'

'He can't give you anything, now. We can't give him anything.'

'You could write a letter to his grandparents.'

'I don't know what to say.'

'Just tell them Thulani was your friend – that you remember him. They'll like that. Write the letter and I'll take it to them when I go back to Muller's Drift.'

'I don't want you to go back to Muller's Drift,' Neil said.

'I must, you know that.'

'Not on my account.' Neil came to stand between Todd and the car. 'I admit that when this thing began, I saw it as a chance to settle with Wivern. All these years it's been eating at me, that he killed Dave and got away with it. But over the past few days I've come to see that that kind of vengeance means nothing compared to your safety. I want you safe, Todd. I don't want to lose another son.'

Todd half-smiled. 'Thanks, but my mind's made up. I'm not doing it for you, or Dave, or Thulani. I'm doing it for myself, something I need to do. You understand?'

'I'll go with you.'

'No, you have to stay and keep in touch with Marx.'

Neil put an arm round Todd's shoulders. 'Just don't take any risks, hear?'

'And you.'

Neil watched the car through the gates, then turned back to the house and went up to Meg's room. Her door was locked. He called her name several times, but she made no answer.

The afternoon dragged. Neil resisted the urge to call Brixton. He dined alone, Meg keeping to her room. At 8.30 the telephone in the study rang.

'Mr Scallan? Marx here. I'd like to have a quiet talk with you, if it's convenient?'

'My place or yours?'

'I think, let's meet on neutral ground. Do you know the Majestic Hotel in Westcliff?'

'Yes.'

'I'll meet you there.'

Neil drove north as fast as the traffic allowed. He found Marx sitting in the shadows of a deserted veranda, most of the hotel's patrons preferring the bustle of the public bar. A waiter was unloading glasses, ice and a siphon on to the table.

'I ordered whisky for you,' Marx said. 'If you'd prefer something else . . .'

'Whisky's fine.' Neil sat down, added soda to his tot and looked at Marx across the raised glass.

'Well, Colonel?'

'You understand this is just a friendly chat, Mr Scallan?'

Neil nodded. Marx sipped his drink and set it aside. 'This morning you raised certain questions. I've obtained some answers. I don't know if you'll like them.'

'Oh? Why's that?'

'The first answer: George Wivern is certainly dead. I've had confirmation of his death from a source I trust absolutely. He was standing beside a fuel dump. There was an MPLA attack, air and ground. Wivern was hit, and fell. The fuel drums exploded and he was incinerated. There were several reliable witnesses. There's no doubt in my mind they gave a true account of events.'

'Then who made the call to me; who sent the note?'

Marx shifted a little and light shone for a moment on his eyes.

'One possibility is that you invented both call and note.'

'I? For God's sake, why would I do such a thing?'

'To get the case of your son's death reopened.'

'No!' Neil shook his head vehemently. 'That's bull. I'm convinced the call and the note were engineered by Frans Oswald.'

'You have no proof of that. Your scenario was based on the premise that Wivern is alive. If he's dead . . . and he is, I assure you . . . your theory falls apart. There's nothing for us to investigate.'

'Then why are we here?' said Neil swiftly. 'Why are we having this quiet talk?'

Marx took his time answering. 'You asked about Ritter,'∙ he said at last. 'You asked if he'd sent in a report. He did. About three months ago, he reported that unmarked vans had been seen moving at night on the back roads in his area. He thought they might be transporting stolen goods. He asked for extra men, and authority to stop and search the vans.'

'Was his request granted?'

'No. With all the violence in Natal, we have to concentrate our men in known trouble spots.'

'Mr Smedley saw an unmarked van the day he was killed.'

'Which doesn't mean the van was connected with his murder.'

'If he saw someone he recognized . . .'

'Wivern, you mean? I told you, Wivern's dead.'

Neil saw that Marx was watching him intently; almost as if he were urging him on.

'Could Frans Oswald have heard of Ritter's latest request?' he asked.

'It would have been highly irregular.'

'But possible?'

'Yes.'

'When Ritter was in Vryheid . . . after the bus incident . . . did he make a report?'

'Of course. He also testified at the inquest. He believed that the White Torch organization had a large cache of illegal weapons. He wanted a search made of the surrounding farms.'

'Was it made?'

'Yes. No cache was found. Ritter continued to insist that arms were being smuggled into the district from Swaziland or Maputo.'

Neil thought for a moment. 'That means through northern Natal. The route would run close to Muller's Drift.'

'Yes.'

'So it's possible that illegal weapons could be stashed there, distributed from there to the Vryheid area, or south to trouble spots like Mooi River and Richmond. The vans Ritter saw – the van Smedley saw – might have been transporting guns. Did you find out if Smedley knew Wivern?'

'I spoke with his daughter, she's a Mrs Schoonbee now. She said her dad never knew any George Wivern. She said he told her everything and he never mentioned Wivern, she'd have remembered a funny name like that. I showed her Wivern's photograph. She didn't recognize it at all.'

'That's hardly conclusive, is it?'

Marx made a tired gesture. Neil saw it was useless to argue further. Marx didn't need theories, he needed facts, especially facts that tallied with those already entered in police records.

Neil was seized with unreasoning anger. 'Right, that's it, then. My theories are moonshine. There's nothing more to say.'

'Wait.' Marx set aside his empty glass. 'Mr Scallan, I didn't ask you here so I could make a fool of you. I asked you because in some respects I agree with your scenario. I've tried to find evidence to support it. I haven't done too well there, but this afternoon I found something very significant.' He laid a photograph on the table, placing it in the shaft of light from the bar. Neil stared at full-face and profile shots of a fair man. The hair was close-cropped, a beard masked the jawline, but nothing could disguise the wide-set incurious eyes.

'That's Leonard Brown,' Neil said. 'Are those police mug-shots, has he a record?'

'His real name is Marius Willem Luther,' Marx said. 'His story about twenty years in the army is bull. He was never in the SADF. He served as a mercenary in the Congo, Zimbabwe and Angola. He's a sniper and demolitions expert. He's suspected of complicity in the murder of several black activists, but there's never been enough evidence for us to bring charges. Luther's a killer, Mr Scallan. He works for anyone who has enough money to pay his fee.'

Neil's mouth was dry. 'You think he killed David?'

'No. I think Wivern did that, but I think it's more than likely that Luther killed Wivern.'

Neil pushed back his chair. 'I have to warn Todd.'

'No.' Marx's fingers closed over Neil's wrist. 'That could be dangerous. If you son shows he suspects Luther, he could get hurt . . . and Luther will certainly clear out.'

'What if he does? I only care about Todd's safety.'

'I've already arranged for extra men to be sent to Muller's Drift, and Ritter knows Luther is at the hotel.'

'I'm going to tell Todd to get the hell out, just leave the bloody place.'

'And will he do as you say?' Marx's tone was ironic, and Neil subsided into his chair.

Marx retrieved the photo from the table.

'Let *me* present a scenario,' he said. 'It begins in 1983. Your son David, engaged in helping refugees across the border, visited Gaborone. There he saw Mark Alberton and George Wivern. He believed they were collaborating to defeat his own operations.

'When he returned to Johannesburg, he told a friend what he had seen, and said he was going to expose Alberton to the authorities. Before he could do so, he was killed by a car bomb.'

'Set by Wivern.'

'Exactly. His killers were never found. The case might have become just another unsolved political murder, except for the determination of two people – yourself, and Frans Oswald. You were convinced Wivern was responsible for the bomb. Oswald was convinced Wivern and his team were part of a ring smuggling arms into the country. Between the two of you, you pinned Wivern in a very bright spotlight. The people he worked with didn't like that, so they removed him from the stage. He was sent to the Caprivi strip on the Angolan border. The unit he joined employed civilian trackers. One of them was Marius Willem Luther. He was in the camp when the MPLA shot it up and the fuel dump exploded.'

'You think he had a hand in Wivern's death?'

'Let's say he might help us with our inquiries. What I think and what I can prove are two different things. But to resume . . . by early 1984, we had the situation that Wivern and Moroka were dead, Silbermann in a geriatric ward, and Alberton exiled to Australia. End of story – except for one thing. Frans Oswald refused to accept the official story of Wivern's death. He continued to accuse senior security people of complicity in gun-running. He became more and more unbalanced. Eventually he was forced to take his retirement.'

Marx took a cheroot from his pocket, eyed it longingly, and put it away again.

'So far, we've stayed in the region of fact, or reasonable supposition,' he said. 'Now we come to plain old-fashioned speculation. Three years ago, after the bus rank massacre, the arms trade was back in the headlines. Sergeant Ritter of the Vryheid police stated that in his view the goods were coming in through northern Natal.

'Oswald thought he'd found a kindred spirit in Ritter. From friends still in the service he learned of Ritter's transfer to Muller's Drift and his report of suspicious activity in that area. Oswald concluded that his theory had been proved right. Wivern was alive, back on the job, working the Natal route.

'Oswald knew it would be a waste of time to appeal to the authorities. Anything he said would be disregarded. He hit on the idea of getting someone else to do his work for him. He hired a man to call you, claiming to be Wivern. He planted a note suggesting that Wivern's gang was paid to assassinate your son.

'He knew he could rely on you to take action. Whether you went to the Press or the police, didn't matter. Either way, the hunt for Wivern would be on. He would be forced out of operation. He might even be arrested, and the ring broken.'

Neil was frowning. 'There are a dozen ways illegal arms come into this country. What made Oswald pick on the northern Natal route? What's to connect that route with Wivern?'

'Criminals are creatures of habit. They return to old associates, old haunts, old methods. Oswald must have picked up a trail he thought he recognized. Unfortunately, it turns out to be a false trail. It leads nowhere.'

'It leads to Muller's Drift,' Neil said. 'To Ritter's suspicions, to the Smedley and Msomi murders. Don't you feel that?'

'I'm not paid to have feelings, meneer. There's nothing to connect these events. When you told me old man Smedley saw someone he knew, I hoped that might provide the link; but now we know Wivern's dead, that theory falls away.' Marx sounded weary. 'I'll see that Ritter gets more men. Beyond that, there's little I can do.'

'We have to find a way,' Neil said doggedly. 'In a few weeks there'll be an election, the most crucial this country has ever faced. There are extremists who've said they will disrupt the polls by violence. They have to be stopped. The people who are supplying them with guns and explosives have to be stopped. Muller's Drift is one of the places they operate. The events there are connected with the arms trade.'

Marx watched him with sombre eyes. 'If you find the connection, Mr Scallan, tell me right away.'

24

The house was ablaze with lights. A cream Rolls-Royce was parked near the front door. Neil knew it belonged to Sadie Morgan, whose husband was on the Energy Commission. As he watched, Sadie and Meg emerged from the house carrying suitcases, which they put in the boot of the Rolls.

Neil went to them.

'Meg? What's going on?'

'I'm moving out.' Her face was pinched with dislike. 'I'm getting a divorce. I'll be staying with Sadie.'

She walked to the passenger door and climbed into the car. Neil looked at Sadie.

'I'll phone her tomorrow. Tell her . . .'

Sadie turned her back on him. She climbed into the car and started the engine. As Neil stepped clear, the Rolls purred down the drive and the security gates closed after it.

'Neil?' Aline came to stand beside him, putting her arm round him.

He said, 'Do you know what this is about?'

'She told me tonight she plans to divorce you and live in the Cape.'

'One of her hysterical fits.'

'No. She means it. She's quite calm.' Aline turned him towards the door. 'Come inside, you're exhausted. It's all been too much.'

'It's not over.' He walked beside her like an automaton.

In the hall she said, 'I'll fetch you some brandy.'

'No. Just leave me, Ally.'

She hesitated, her eyes full of concern.

'Please,' he said.

She touched a hand to his cheek and went away to her own part of the house.

Neil walked into his study and flopped down in an easy chair. He felt disoriented. The furniture and pictures looked unfamiliar, the distances between them somehow distorted. He knew he must try to think, but the effort was too great. He closed his eyes and sat quite still for what seemed a long time. At last an idea floated to the surface of his mind. He reached for the telephone and dialled the number of the Muller's Drift Hotel. Presently Todd came on the line.

'Todd?'

'Yes. Dad? What is it, what's happened?'

'Your mother walked out.'

'What? This is a bloody awful line. What did you say?'

'Your mother wants a divorce.'

'Oh.' Todd sounded relieved if anything. 'It's not exactly a surprise, is it?'

'I suppose not.'

'Are you all right?'

'Tired. I'm coming to join you, tomorrow.'

'Good, then. Good. Has something . . . ?'

'No. I don't know quite when I'll arrive, but wait for me at the hotel. Don't go out. Understand?'

'Yes, sure. I'll work on my specimens.'

'I'll see you tomorrow.'

Neil hung up. It was too much trouble to leave his chair, so he stayed where he was. The light of the single lamp shone in his eyes so he switched it off. He allowed himself to drift.

He woke hours later, cramped and cold. The inertia of shock had worn off. He felt calm and clearheaded.

He knew what had troubled him about Marx; not anything

that had been said, but the whole pattern of Marx's behaviour over the past few days. That was what jarred.

Marx had acted out of character. He was obviously a man to whom correctness mattered, yet he had not followed correct police procedure. He had acted very much on his own, to the point of arranging tonight's one-to-one discussion. He had broken rules, divulged information that would normally remain secret.

It was inconceivable that a man who had built a reputation for intelligence and discretion should suddenly become both stupid and indiscreet.

It followed that Marx had said and done exactly what he'd planned. He had fed Neil certain information. He had made it clear that the police would not continue their investigation unless they found fresh and sufficient grounds to do so.

Significantly – very significantly – when Neil had talked of pulling Todd out of Muller's Drift, Marx had argued against it. He wanted the inquiry to continue, despite any danger to Todd. He wanted to test the scenario.

If Marx had acted without authority, then he could be that most dangerous of animals, the rogue cop.

If he had acted on orders, then one must ask who gave those orders, and why.

Did his shadowy backers want to expose old crimes, break an arms racket? Or did they intend to complete the cover-up of the past and silence any challenger?

It was nearly morning, the sky already pale above the black shapes of the trees. He needed a bath and shave and something to eat. He was on his way to the door when the telephone rang. He turned back to answer it.

'Neil?' Louis Baird sounded cheerful. 'Did I wake you? It's afternoon here. I'm sorry.'

'Where are you speaking from?'

'Hong Kong. I have news for you about Mark Alberton.

What made you think he was killed in a road accident?'

'He was. His father told me he was run down.'

'By a ship, not a car. Apparently . . .'

'Wait.' Neil reached to turn on the message-recorder. 'All right. Go on.'

'Alberton worked for a company operating small boats at a place called Keppel Bay, up the coast from Brisbane. They did commercial fishing, sometimes ran a wealthy tourist out to look at the Great Barrier Reef, that sort of thing.

'Alberton was a good seaman, but a bad risk in other respects. He kept company with a bunch of crooks in Brisbane, people the Oz police thought might be bringing in drugs. They had their eye on Alberton, but before they could move, he did. He made a night-run in one of the company boats, possibly to keep a rendezvous with a larger craft. He never came back to shore.

'His folks reported him missing, and a search plane found the wreckage of his boat. Alberton's body was not found, but there are sharks in that area, so the verdict was he went the way of all flesh.'

'You're sure about this, Louis?'

'Yuh. Had it from the Brisbane Chief of Police, a very sharp citizen.'

'Thanks.' Neil's thoughts were racing. 'Thanks, Louis, I'll call you back sometime. Right now I have things to see to.'

It was just on seven o'clock when Neil arrived in Irvine Crescent. At Number sixteen, Mr Mercer was already up, wrapped in a woollen dressing gown and examining his rose bushes. He greeted Neil with a raised hand.

'Mornin', Mr Scallan. You're up early.'

'So are you.' Neil climbed out of the car and approached the gate.

'Rose beetles.' Mercer displayed a tin can full of dead

insects. 'You have to pick 'em off by hand while it's still dark. What can I do for you?'

Neil entered the garden, closing the gate after him.

'I need some information.'

''Bout what?'

'Did you ever know a Mr and Mrs Albert Smedley?'

Mr Mercer chewed his upper lip. His peppery expression sharpened.

'It's a common name, Smedley. I knew some. Why?'

'A man of that name and his wife were murdered a few days ago.'

'Murdered?' Mercer went pale. 'Where? How?'

'At a place called Muller's Drift, in Natal. They were farming there. Mr Smedley was shot. His wife's throat was cut.'

'My God!' Mercer backed towards a rustic bench and sat down, lowering the can of beetles to the grass. He stared at Neil and said again, 'My God!'

'You knew them, sir?'

The old man shook his head. 'Not really. I met them a couple of times.'

'Where?'

'Right here.' Mercer stabbed a stained finger at the ground. 'I bought this house from their daughter's husband. Dolf Schoonbee married Smedley's daughter Lorraine. Dolf's a friend of mine. When I took early retirement – ticker trouble – Dolf took over my house and I bought his. It's a good investment, this part of town's got flat rights, I've been offered—'

Neil interrupted the flow. 'Did Lorraine Schoonbee's parents visit her here?'

'More'n visit, they lived here quite a few months. Bert Smedley was retrenched. He wanted to go farming, put his severance money into a place in Natal, but before he left Joeys him and his wife stayed here with Lorraine.'

'Was the farm at Muller's Drift, Mr Mercer?'

'That I can't say. I only met them a couple of times, never knew what their plans were.'

'When did they leave Johannesburg?'

Mercer pondered. 'That'd be . . . 1980 or so. I bought from Dolf in 1981.' He peered up at Neil. 'What a terrible thing, murdered! There's too many these days. Old folks aren't safe, specially on farms. I hope they get the bastards that did it.'

'Can you give me Mrs Schoonbee's present address?'

'It's nine Tollman's Grove, Linden. They're in the book under A.L. Schoonbee. A for Adolf.' Mercer was staring at Neil. 'How come you're asking all these questions? You a friend of the Smedleys?'

'My son knew them. Thank you, Mr Mercer, you've been most helpful.' Neil retreated through the gate, climbed into his car, waved a farewell.

Mr Mercer continued to sit on the bench. He eyed the dead beetles in the tin can.

'That's what they should do to the murdering buggers,' he muttered. 'Finish 'em off, quick.'

In his office Neil spent an hour with a tape recorder, setting out the events of the past week. He gave the completed tape to his secretary and told her to take it at once to his attorneys, Brice and Thompson, whose offices were in the same building.

Then he called Colonel Marx and fixed an immediate appointment. On his way to Brixton he stopped at a shopping mall and called Anna. When she answered, he said, 'Are you alone?'

'Yes.'

'I have to go out of town for a bit. I wanted to tell you: Meg wants a divorce. I'll be free to marry you, if you'll have me.'

'Of course I will.' She sounded between tears and laughter. 'But can't you come over, can't we talk?'

'No, I have to leave at once, to join Todd. I won't be away long. Tomorrow we'll talk. Tomorrow, OK?'

'Yes.'

'I love you.'

'Yes. Love you.'

'If you have any problems, go to Tommy Brice. He'll sort things out.'

'Right. Neil? Please take care.'

Leaving the mall he passed a florist's shop. He bought a basket of spring flowers and asked that they be delivered to Anna's flat.

Marx and Joubert received him together. Neil told them of his interview with Mr Mercer.

'If Bert Smedley stayed in Irvine Crescent with his daughter, during 1980,' Neil said, 'they must have met the Albertons who lived next door. Mark Alberton lived with his parents; he was never in a varsity residence, I checked. Talk to Lorraine Schoonbee, ask her if her father knew Mark Alberton.'

Joubert said quickly, 'Even if he did, Mr Scallan, it makes no difference. The boy's dead. He was lost at sea.'

So they'd known that all along. Neil looked at Marx's impassive face.

'The lost can be found,' he said. 'Did you know that Mark Alberton was about to be questioned by the Australian police when he so conveniently disappeared?'

Marx made no comment, and Neil continued, 'If he was using his boat to run drugs, he must have had offshore suppliers, probably on ships from the Far East. He could have used those contacts to help him get clear.'

Marx sighed. 'Mr Scallan, it won't help to present us with another set of wild guesses.'

Neil held Marx's gaze.

'Do you play bridge, Colonel?'

'Sometimes.'

'Have you heard of Murphy's Option?'

'I'm afraid not.'

'There are times,' Neil said, 'when the only chance a player has of making his call is if certain key cards lie in the right places. In that situation he must play the hand as if the cards lay that way. He must take Murphy's Option. That's what we have to do now.'

Joubert started to speak but Marx silenced him with a gesture.

'We have to assume that the cards lie in a certain way,' Neil said. 'We have to assume that certain facts obtain. Alberton was involved with Wivern in gun-running operations. When he was forced to emigrate to Australia, he took up with a criminal gang. When it looked as if the police were catching up with him, he planned his vanishing act. He fixed a night rendezvous out at sea with some of his pals. He was picked up by them, and either destroyed his own boat, or left it to drift in the sea lanes. He travelled to one of the dragon states, and continued his unlawful activities. He joined a gun-running cabal, and in due course returned to Africa – he was suited to the job because he knew the local languages, had worked with Wivern and knew his routes. He set up his operation in northern Natal, using Wivern's agents and safe houses. He stashed arms in Muller's Drift.

'For a time, things went well. Then Sergeant Ritter was posted to the area, a man with a reputation for hunting down arms dealers. Frans Oswald also picked up what he took to be Wivern's trail and began to make waves.

'Alberton became nervous. One night when he was

moving goods, his van broke down near Smedley's farm. Bert Smedley nosed around and recognized him. That night Smedley and his wife were murdered.

'A few nights later, Thulani and Siphiwe Msomi went out collecting herbs. They ran into Alberton's thugs, and were killed.

'Alberton is a badly scared man. The chances are he will try to move his stock of weapons to some safer place. If he does, we can and must catch him at it.'

Marx sniffed. 'Having made all these assumptions, Mr Scallan, how do you propose to play your hand?'

'I don't,' Neil answered. 'I'm the dummy. I put my cards on the table. You play the hand.'

25

'I'm not going out today,' Todd said. 'My dad's coming down for the inquest. I want to be here when he arrives. Besides, it looks as if there's going to be a helluva storm.'

The fair man squinted at the sky. 'Maybe later.' He fingered an ear. 'Your dad's a lawyer, isn't he? I heard the name somewhere.'

'He's an advocate.'

'He's going to act for the Msomis?'

'A watching brief, perhaps.'

The fair man looked unconvinced. He seemed uneasy as an animal scenting lion. 'I'll be here if you want me,' he said, and perched on the veranda rail, staring moodily at the street.

Todd fetched the letter Joe had written to Hosea and Ellen, and found an umfaan to take it up to the Msomi kraal.

The rest of the morning he spent working in his room, sorting and fixing specimens, and writing notes.

Neil arrived soon after three o'clock. He dumped his bag in Todd's room and said, 'We need a place to talk in private. I've a lot to tell you.'

'The garden,' Todd said. 'No one goes there.'

They walked between overgrown flowerbeds and lawns parched with drought to a bench near the back gate. Neil recounted all that had happened since their talk with Marx and Joubert. Todd listened without comment and at the end said, 'So the action's here. What do we do?'

'Our job is to beat the covers; scare Alberton into moving whatever he has hidden in this area. The rest is up to Marx. He has to persuade the security forces to cordon off the valley and prevent any getaway.'

'You think Marx will do that? You trust him?'

'I believe he's an honest cop. He'll weigh the odds carefully. He won't send his men on a wild-goose chase, but in the final analysis he'll act to break the arms-runners. He'll back us.'

Todd picked a dead head of lavender and rolled it between his fingers, releasing its pungent scent.

'Does Ritter know what's planned?'

'Marx will brief him.'

'When do we start?'

'Tomorrow, at the Msomi inquest. Everyone will be in the village, if not in the courtroom. You'll be called to testify. We'll work out what you must say. It'll be enough to frighten Alberton.'

'I hardly think he'll be in court:'

'He'll get word fast enough. My bet is, he'll arrange to move the arms out of the area.'

'If there are any.'

'There are. Murphy's Option.'

'It's risky. These people are killers. If Marx lets us down, we're in big trouble.'

Neil glanced back at the hotel.

'I'm thinking of calling in a hired gun,' he said.

They were drinking their after-dinner coffee in the lounge when a waiter brought Todd a package wrapped in brown paper.

'From up there,' he said, indicating the hill behind the hotel.

Todd unwrapped the parcel. It contained a few sheets of

greaseproof paper, covered with pencil tracings. There was also a note from Ellen.

Dear Mr Todd,
I found these in Thulani's room, I do not know what they mean, I send them to you.

'*Encephalartos woodii*,' Todd said. He showed the drawings to Neil. 'Thulani and I were talking about cycads. I told him how rare *woodii* is, that there are no female plants in existence.' He felt tears burn his eyes. 'He had cycad seeds in his belt-pouch when he was found. That must have been the present.'

'What present?'

'Joe told me Thulani wanted to give me something. He must have been looking for the cycad when he was killed.'

He stared at the tracings, so carefully executed: seed-cone, rachis, leaflet.

'Where did he make those?' Neil asked.

'I suppose at our place.' Todd had a sudden image of the book Thulani must have used. He visualized the illustrations, the text, the cover of the book that showed in bright colour a cycad in cone, with birds feeding on the seeds.

'Hornbills,' he said. 'Trumpeter hornbills.' He heard their harsh cries, saw them tumbling about a burning sky.

'I know where the Msomis were killed,' he said.

The inquest was held in the village hall, at 3.00 the next afternoon.

The Msomi clan was there in force, the elders, in formal clothes, ranged near the front; the women and young people in the rear gallery. The white population was also well represented. Todd saw Kramer sitting with the stationmaster, and du Toit just behind them. Sergeant Ritter nodded a greeting, but did not approach. Luther was there, sitting near the main doorway.

The coroner, a small balding man in steel-rimmed spectacles, gave Neil a sharp glance as he took his place, and Neil wondered if they had met before. Ritter would have warned him that Todd wished to speak.

The proceedings moved slowly, as much of the Khumalo testimony was given in Zulu and had to be translated. Ritter set out the circumstances of the finding of the bodies in the gully on Khumalo territory. The district surgeon who had performed the autopsy confirmed that the victims had died of knife and panga wounds. A Khumalo headman denied with passion that any member of his tribe had had anything to do with the crime. Hosea Msomi stated that his brother and grandson would never have trespassed on Khumalo land. 'They would have stayed on our ground,' he insisted. 'Mr Scallan knows I speak the truth. My grandson was carrying seeds from plants that do not grow on Khumalo ground. He must have gathered them on our land.'

'You mean, on Msomi land,' said the coroner, and Hosea nodded.

'Yes, sir. On Msomi land.'

Todd was called to the stand. He identified himself and gave his qualifications as a botanist.

'I understand,' said the coroner, 'that you wish to speak in support of what Mr Hosea Msomi has told us?'

'Yes, your honour.'

'Did you see the seeds referred to by Mr Msomi – the ones found in the belt-pouch of Thulani Msomi at the time of his death?'

'I did.'

'Did you examine these seeds?'

'Yes, in the office of Sergeant Ritter. They were cycad seeds from the female cone. In my view, they had been recently

181

gathered. There was damp mast and loam clinging to them.'
Todd hesitated, and the coroner nodded encouragingly.

'I believe,' Todd said, 'that Thulani's purpose in going out
with his uncle on the night he died, was to try to locate an
Encephalartos woodii cycad, an extremely rare plant. Thulani
spent a weekend at my home in Johannesburg a short time
ago. He told my son he wished to make me a present. He
searched through one of my books on cycads, and made
tracings of the *woodii*. Those tracings were found among his
effects by his grandmother. She sent them to me yesterday.'

'Mr Scallan, would this cycad – *woodii* – be found on
Khumalo ground?'

'No. It's too dry and exposed. *Encephalartos woodii*'s natural
habitat is in afforested regions.'

'Do such regions exist in Muller's Drift?'

'Oh yes. There's natural forest on the southern boundary
of Albert Smedley's farm, and also in one or two places near
Crown Hill. Thulani knew this. We spoke of it at length.'

The coroner consulted his notes. 'Mr Scallan, we have
heard Mr Hosea Msomi state that no member of his family
would have gone herb-gathering beyond the limits of Msomi
territory.'

'That's true, but you have to remember that Siphiwe
Msomi was an old man and a traditionalist. He was born and
raised on the highlands near Crown Ridge. That was all
Msomi land at the time. Twenty years ago the tribe was
forced to move to where they now live, north of the village;
but to Siphiwe, Crown Ridge and its environs were the
property of his people. He would not have considered it
trespassing to gather herbs there.'

'Are you suggesting that Siphiwe and Thulani Msomi may
have been murdered on the west side of the village?'

'Yes, I am.'

The coroner glanced at Sergeant Ritter. 'No doubt the

police will keep that in mind as they pursue their investigations.'

A verdict of murder by person or persons unknown was returned, and the court rose.

As people began to move towards the exit, Todd saw Fanie Kramer thrusting through them. He followed, and was in time to see Kramer jump into his truck and drive away at speed towards the main road.

Luther was waiting for them on the steps of the hotel, his expression baleful. He blocked Todd's path.

'What did you mean back there?' he demanded.

'What I said; that I believe the Msomis were killed in wooded territory on the west or northwest side of the village.'

Luther shifted his gaze to Neil. 'What are you after? What are you doing here?'

'That's hardly your business, Mr Luther.'

Luther's face froze. 'My name's Brown.'

'Your name is Marius Willem Luther. The police in Johannesburg identified you, and gave me a précis of your record. They're anxious to talk to you. They think you may be able to help them with their inquiries.'

'Inquiries into what?'

'Well, for one thing they'd like to discuss the phone call you made to me, and the note you left for me at Jan Smuts Airport. To be honest, I don't think they're worried about my hurt feelings; but they are very interested in your reason for impersonating George Wivern. His resurrection from the grave really does bother them.'

Luther made a derisive sound. 'Wivern's dead.'

'You should know, of course. Did you also know that Mark Alberton is alive and well and running guns through this valley? That's the trail your pal Frans Oswald picked up.'

Luther said nothing, but his face seemed to shrink. His eyes flickered towards the surrounding country.

'I wouldn't consider it,' Neil said cheerfully. 'There are only three roads out of here: the main road, the Pomona road, and the road south. They're all being watched. To be caught on the run wouldn't help your case. On the other hand, if you stay here and do as I tell you, it might pay dividends.' Neil saw the greedy glint in Luther's eyes and shook his head. 'Not in cash — I've no intention of hiring you, but I'm prepared to give you some free advice.'

Luther licked his lips. 'All right. What?'

'At the proper time, give yourself up to the police. Admit your past crimes, make a full confession and apply for indemnity under the Amnesty Act. You'll have to co-operate, of course . . . tell them where Oswald is, admit you worked for him and posed as Wivern on his instructions. You'll have to tell them you helped Wivern out of this world. Tell them who hired you to do that.'

Luther was white. 'You crazy? I wouldn't last long . . .'

'You won't last long if you try and go it alone. Do as I say, and you have a chance. I have some influence in legal circles, and with the police.'

Luther hesitated, still staring at the hills. Then he reached a decision.

'What do you want me to do?' he said.

At six o'clock the hotel bar was crowded. Todd bought a beer and carried it out to the veranda. Presently Ritter came up the steps from the roadway.

'Colonel Marx phoned,' he said in a low voice. 'Everything's set.'

'Good. Is Kramer back?'

'Ja, but there's nothing moving over at his farm.'

'They'll wait for dark.'

184

Ritter nodded. 'I've got men posted. Anything starts, I'll know, and I'll send the signal. You stay in the hotel, Mr Scallan. There can be shooting.'

Todd didn't answer, and Ritter, taking his silence for consent, hurried away towards the police station.

Todd set his beer mug down and strolled to the end of the veranda, dropped to the path and made his way quickly to the back gate. His station-wagon was parked there, with Luther at the wheel and Neil in the passenger seat. Todd climbed into the back, picking up the rifle that lay there.

'All right. Let's go.'

They drove along the rough track, reached the tar, crossed over it and began the ascent to Crown Ridge.

Luther said, "This is dumb. They'll see us."

'That's the idea,' Neil answered. 'Can you imagine anything more likely to draw Alberton's attention, than the three of us?'

Luther blinked, then surprisingly laughed. 'Yeah. Slug bait. That's what the old toad wanted all along.'

'Oswald?'

But Luther shook his head, refusing to be drawn.

Neil said over his shoulder to Todd, 'Is anyone following?'

'No.'

They were on the new track now, that led to the abandoned dam works. On Todd's order, Luther parked the vehicle in a fold of ground where it would not be visible from below. They climbed out and walked up a grassy knoll, using a footpath that must once have led to a cattle kraal. An outcrop of rock offered natural cover, and they settled behind it.

Far to the east, towards the invisible coastline, they saw thunderclouds boiling, edged with turmeric fire. From time to time lightning flared, but they could hear no thunder yet. Above the plain the moon soared like a balloon rising on the hot night air, illuminating in sharp focus the drought-bleached veld, the river shining like a snail's track, and the glow of farm settlements. Kramer's house showed lights, but the yards round it seemed deserted.

To their right was the raw red earth of the dam-cut, and to their left, the dry bed of the stream that led to the head of the gorge, the haunt of trumpeter hornbills, the place where Siphiwe and Thulani had died.

Todd handed his rifle to Neil, and studied the plain through field glasses.

'Quiet as hell,' he said.

'Have to wait,' Neil said.

Luther ignored them, rolling on to his belly, finding a position that gave him a clear view of the slopes below.

Hours passed. They spoke little, each absorbed in his own thoughts; each wondering if there were men posted on the exit roads.

At a little past midnight they caught the thrumming of engines in the distance, and saw two vehicles travelling slowly along the road from the village: closed vans, using only their dimmed headlamps. They passed Kramer's gate and veered west towards the gorge. In a short while they were lost to view, hidden by the lip of the escarpment. Todd started to edge forward, but Neil caught his arm and held him back.

'Keep still. They have to load up. They'll be back.'

Ten minutes passed. Twenty. The lightning was much closer, now, the thunder rolled, the storm devoured the low country and raced for the hills. They could feel the first gusts of its hot breath on their faces.

'They're coming back,' Todd said. 'They're going to take the south road.'

The two vans lurched into sight, heading across the plain towards the river. Behind them came a jeep. Todd focused the binoculars and swore.

'I can't see faces.'

Neil punched the ground. 'Christ! Where's Marx? If he's ratted on us . . .'

The thunder rolled again, and behind it they heard a faint droning.

'Chopper,' Neil said. He looked south, scanning the sky. 'I can't see it. I can't . . .'

The three vehicles had reached the bridge. Slowly, the first van crossed over, then the second. As it did so, the droning exploded into a roar, and an army helicopter burst into view, veered, bore down on the vans. They put on speed, evidently trying to reach the cover of the hills. There was a rattle of automatic fire. The first van kept going, the second spun off the road and rolled.

The jeep had not crossed the bridge. Instead it reversed, checked, then swung off the road and headed across the veld towards the dam-cut. The chopper, busy with the vans, did not appear to notice.

'He can't drive up this ridge,' Todd said.

Neil snatched up the binoculars and focused on the jeep. It reached the bottom of the slope and slewed to a halt. The driver jumped out and began to scramble up the cut, passed the rudimentary dam-wall and reached the river bed. His right hand held a machine-pistol, his left clawed for hand-holds. They could see his face now, pallid, gasping open-mouthed.

'It's Alberton,' Neil said, and stood up.

The man below caught the movement and for a moment was motionless, staring at Neil. Then his right arm lifted and swung in an arc. As Todd pulled Neil down, bullets sprayed the rock face. Luther fired twice. They heard Alberton scream, and the sound of his gun bouncing down the slope. Then there was silence.

Luther crawled crabwise to where Alberton lay. Neil and Todd followed.

'Is he dead?' Todd said.

Luther didn't trouble to answer. Alberton's throat was torn out, and blood from the hole in his chest ran into the shallow pool beneath him. Luther crouched over the body like a predator guarding his kill. His eyes shone up at Neil.

'You remember what you promised? I did what you said.'

Neil nodded. Down on the plain the vans had been surrounded. They looked like beetles about to be devoured by ants. On Kramer's farm, floodlights sprang up and sirens wailed.

There were sirens on the ridge too, racing towards them along the high contour road.

Todd had taken out a torch and was signalling to the chopper. It flew towards them, banked and hovered. The loudspeaker crackled instructions. Todd waved acknowledgement and went to sit on a ledge of rock.

Behind them two vehicles appeared and spilled out men in camouflage uniform. Sergeant Ritter ran past them and scrambled to where Mark Alberton lay. He stooped over the body for a short while, then came back to Neil.

'You know him?'

'Yes. He was Mark Alberton.'

'Who shot him?'

'I did,' Luther said. He held out his rifle. 'I want to make a statement.'

Ritter took the gun. 'At the station,' he said. He signed to two of the uniformed men who moved in to flank Luther.

Neil said, 'Sergeant, Mr Luther wishes to apply for indemnity. He has important evidence to submit. I suggest you call Colonel Marx right away.'

188

Ritter nodded. 'You coming down, Mr Scallan?'

'In a little while, yes.'

The policemen and Luther moved away. Others approached Alberton's body and began the routine task of taking photographs and measurements. Neil went to sit beside Todd.

The thunder was almost ceaseless now, and the first raindrops fell hissing on the earth, releasing a herbal incense.

Neil thought, this is our time of amnesty, the crimes exposed and the judgement handed down. Later must come acceptance and forgiveness for Meg and Anna and me, for the Msomis and old man Alberton and all the victims of the past; but the terrible years of desolation that so nearly destroyed us all, those at least are over.